Sir Alexander Ogston, 1844–1929

Sir Alexander Ogston, 1844–1929

A Life at Medical and Military Frontlines

David A. Rennie

EDINBURGH
University Press

Edinburgh University Press is one of the leading university presses in the UK. We publish academic books and journals in our selected subject areas across the humanities and social sciences, combining cutting-edge scholarship with high editorial and production values to produce academic works of lasting importance. For more information visit our website: edinburghuniversitypress.com

Edinburgh University Press Ltd
13 Infirmary Street,
Edinburgh, EH1 1LT

First published in hardback by Edinburgh University Press 2024

Typeset in 10.5/13pt Sabon by
Manila Typesetting Company, and
printed and bound by CPI Group (UK) Ltd, Croydon, CR0 4YY

A CIP record for this book is available from the British Library

ISBN 978 1 3995 0131 6 (hardback)
ISBN 978 1 3995 0132 3 (paperback)
ISBN 978 1 3995 0133 0 (webready PDF)
ISBN 978 1 3995 0134 7 (epub)

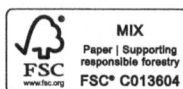

MIX
Paper | Supporting
responsible forestry
FSC
www.fsc.org   FSC® C013604

Contents

Figures

Acknowledgements

I would like to thank the entire Ogston family for the encouragement, support and – importantly – trust they offered me as I wrote this biography of their hugely accomplished and influential forebear. During the summer of 2022, it was my pleasure to meet several of Ogston's great-grandchildren in Aberdeen, where we gathered to attend the official unveiling of the plaque commemorating Sir Alexander Ogston at 252 Union Street. The following day, I joined some of Ogston's descendants in journeying out to his former country home, Glendavan, in Deeside, Aberdeenshire, where we visited Moira Milne – a retired GP and previous owner of Glendavan. The literal travelling and gathering together of the aforementioned signified, to me, the wider reality that the project has, in a sense, been a collaboration between many individuals who have served as curators of Ogston's legacy or as co-conspirators in getting this book into print.

For several years, I have carried on an enjoyable correspondence with Andy Philpot, who deposited Ogston's papers at Aberdeen University, where they were archived by Paul Logie. Both Andy and Katherine Paine provided invaluable assistance in transcribing Ogston's journals. David Woodman allowed me to visit Glendavan, and Moira kindly shared her recollections concerning the house (including what it was like to run it as a lavish B&B), as well as reading the manuscript of this book. In that regard, I am also grateful to Dr Tom Scotland, Dr Grahame Howard, Dr Jane Bedborough, Andy Philpot, and Katherine Paine. Additionally, I would like to acknowledge the assistance provided by the following institutions and individuals: Aberdeen Archives, Gallery and Museums; Aberdeen University's Museums and Special Collections; Allison Derrett, Archivist at the Royal Archives, Windsor; Alastair Fiddes Watt; and the Wellcome Trust, London.

Introduction

SURGERY UNDERWENT SEVERAL CRUCIAL developments during the life-time of Sir Alexander Ogston (1844–1929). At the time of Ogston's birth, operations were performed without anaesthesia, often with the patient held in place on the operating table by the physical strength of the surgeon's assistants. Moreover, antiseptic methods were neither under-stood nor implemented. Even if the patient survived a surgical procedure, therefore, they were at risk of fatal post-operative infection. Warfare, also, would change dramatically during these years. Nineteenth-century cavalry manoeuvres and square-formation marching were replaced, in the twentieth century, by the modern, impersonal warfare of machine guns, poison gas, and high-explosive shells.

Ogston was not merely an observer of these changes. His roles as Senior Surgeon at Aberdeen Royal Infirmary and Regius Professor of Surgery at Aberdeen University were enough, alone, to constitute a highly distinguished career. Ogston's commitment to his profession, however, was such he could not overlook opportunities to refine, develop, and encourage improvements in contemporary medicine. On the operating table, in the laboratory, or at the front, he demonstrably advanced civil-ian and military medical practices throughout the course of his long and remarkable life.

Aseptic methods opened possibilities for elective surgery, and Ogston would pioneer important new operations for *genu valgum*, club foot, and flat foot. Moreover, inspired by Robert Koch's work in the nascent field of bacteriology, Ogston was prompted to ask what, exactly, was the cause of post-operative suppuration, inflammation, and blood poisoning? His research into this question was to yield profound and far-reaching results. Ogston conclusively demonstrated acute inflammation and sup-puration were caused by microorganisms; he discovered (and named)

Staphylococcus; and Ogston correctly linked localised microorganism infections with blood poisoning. In 1884, Julius Rosenbach identified different forms of *Staphylococcus*, which he named *Staphylococcus albus* and *Staphylococcus aureus*. During the 1960s, the latter strain became resistant to penicillin. Its resistance to antibiotics makes methicillin-resistant *Staphylococcus aureus* (MRSA) challenging to treat. Furthermore, as they may have weakened immune systems, open wounds, or may be fitted with a catheter or intravenous drip, hospital patients are more susceptible to MRSA infection – thus, MRSA is also known as hospital 'superbug'.

Being a renowned surgeon residing in the north-east of Scotland were inducement enough for Ogston to be offered, in 1892, the post of Surgeon-in-Ordinary in Scotland to Queen Victoria. And Victoria's approval would prove valuable to Ogston as he pursued another key personal endeavour: improving British military medicine. Having volunteered in the Soudan (Sudan) war in 1885, Ogston was quick to realise the infrastructure of the army medical services (and the training of its personnel) had fallen behind advances made elsewhere in medicine and was inadequate for its ostensible purposes. Ogston formed part of the redoubtable medical delegation that met Lord Lansdowne, Secretary of State for War, in 1898, resulting in the formation of the Royal Army Medical Corps that year. Ogston's belief that the British Army Medical Service was unprepared for a large-scale conflict was amply confirmed during the Boer War, in which Ogston voluntarily – and supported by Queen Victoria's official sanction – offered his medical assistance to Lord Methuen's forces. Later, during the Great War – upon the eve of which Ogston's presidency of the British Medical Association began – he was, again, keen to offer his skills. Ogston served with distinction at the Villa Trento hospital in north-east Italy, a site which served, in part, as the real-life inspiration for the British hospital where Catherine Barkley and Frederic Henry meet in Ernest Hemingway's *A Farewell to Arms* (1929).

A life of such indisputable interest and significance is not only a story worth telling in full, but one that merits fuller appreciation than it has yet received. Ogston's comparative obscurity emanates, perhaps, from the diversity of his career. He was a surgeon at Aberdeen Royal Infirmary, a Professor at Aberdeen University, a bacteriologist, a royal surgeon, a co-founder of the Royal Army Medical Corps, a surgical pioneer, and an individual who volunteered to serve in the field during three wars. The wide range of these accomplishments may have precluded celebration of their cumulative distinction. Ogston cannot be held up primarily in association with a single field of accomplishment – as Sir James McGrigor can, for instance, in relation to his role as a superlative medical administrator. Rather, Ogston was an individual of variegated interests and accomplishments. For the first time, this book weaves the separate strands of Ogston's excellence into a single narrative of his life's work, one for which the fields of bacteriology, surgery, and military medicine were substantially enriched.

1

1844–1873

I saw that a miraculous change had come over our Science, and my mind was almost bewildered with the glorious visions of all that it entailed. I felt inclined to sit down, cover my face with my hands, and think out what the great revelation implied in the future.[1]

IT WAS CHARACTERISTIC OF Ogston that, when the British Medical Association gathered in Aberdeen for its annual conference in 1914, he chose the history of Aberdeen University's medical faculty as the topic of his presidential address. As well as being an eminent surgeon, reformer of military medicine, and pioneering researcher, Ogston had a wide range of interests he pursued with the diligence and exactitude that shaped his professional career. Ogston was multilingual, a genealogist, an archaeological researcher, a photographer, a man who loved to shoot game on his Deeside estate. And his 1914 address to the BMA reveals an intellectual life of wide range, capacious understanding, and scrupulous attention to detail.

Although titled 'On the Making of a Scottish Medical School', it is more accurate to describe Ogston's presentation as a condensed history of north-east Scotland. He began his address – delivered at Aberdeen Music Hall on 28 July – by describing the autochthonous Pictish peoples of the north-east of Scotland and the influx of Romans, Anglo-Saxons, and Vikings. Ogston then outlined the topography a visitor to Aberdeen would have encountered in the late fifteenth century. Continuing north after fording the River Dee, the traveller would reach 'the valley where stands the railway station at which you [his audience] arrived now stands, where the ground was then a quagmire, partially covered with mud and seaweed, and partly a reedy expanse where the wild-fowl bred in summer'.[2] Ogston related how William Elphinstone (1431–1514), having been made Bishop of Aberdeen in 1483, arrived at this 'district inhabited by ignorant and nearly savage men' and set about establishing a university.[3] In 1494 Elphinstone obtained, via a petition from James IV, a papal bull from Pope Alexander VI granting permission for such an institution. This made Aberdeen the third university in Scotland – coming after St Andrews (1413) and Glasgow (1451) – and the fifth in Britain.[4]

Ogston informed the delegates, thanks to Elphinstone, Aberdeen University could also claim to have 'the oldest Faculty of Medicine in any university in Great Britain or Ireland'.[5] Elphinstone's 'clear vision saw, what most of his contemporaries did not discern, that medicine, like other branches of knowledge, was likely to share the impetus resulting from the' Renaissance's renewed interest in Greek literature – including the medical writings of Hippocrates, Aretaeus, and Galen. Elphinstone also obtained from King James a charter gifting 'to a graduated Doctor in the faculty of medicine, the sum of £12 6s. yearly'.[6] And James Cuming was the first 'Mediciner' or Professor of Medicine. Ogston positions Elphinstone, therefore, as a medical reformer – someone prepared to make exertions to ensure standards evolved to meet changing contemporary circumstances. Ogston would follow such a path himself. Indeed, while he would probably shrink from conceiving of himself in such terms, Ogston's career places him among a long lineage of medical notables with attachments to the University of Aberdeen. These include, for instance, George Cheyne (1671–1743), an early advocate of vegetarianism and author of *An Essay of Health and Long Life* (1724); Sir James McGrigor (1771–1858), who between 1812 and 1814 served as Inspector-General of Hospitals for the Duke of Wellington's army in the Peninsular War (1808–14) and Director-General of the Army Medical Service (1815–51); Sir James Clark (1788–1870), meanwhile, authored *The Influence of Climate in the Prevention and Cure of Chronic Diseases* (1829) and was Physician-in-Ordinary to Queen Victoria (1837–1860). And Ogston's father, Francis (1803–87), served as Professor of Medical Jurisprudence at Aberdeen University. Ogston's willingness to, where necessary, lead rather than merely follow his profession places him alongside such individuals in the pantheon of British and global medicine.

Ogston was a keen researcher of his family history and authored two privately printed books on the subject: *A Genealogical History of the Families of Ogston from their First Appearance circa. A.D. 1200* (1876) and *Supplement to the Genealogical History of the Families of Ogston* (1897). Ogston traced his lineage as far back as Symon de Hogeston (died before 1240) of Morayshire. Among the children of his paternal grandfather, Alexander Ogston (1766–1838) – a manufacturer of soap and candles – were the sons Alexander Ogston of Ardoe (1799–1869) and Francis Ogston, MD. Alexander Ogston of Ardoe continued his father's manufacturing business, making soap and candles on a premise in Aberdeen's Gallowgate. So prosperous was this family enterprise, 'Soapy Ogston' became a byword for wealth in Aberdeen.[7]

Alexander earned the 'of Ardoe' appendage after purchasing Ardoe estate in 1839.[8] Of his offspring – Sir Alexander's cousins – two continued the family business. Alexander Milne Ogston (1836–1926) bought the Ardoe lands from his father's trustees in 1870 and commissioned Ardoe House, built in Scottish baronial style in 1878.[9] His brother, Colonel James Ogston (1845–1931), purchased Norwood Hall at Cults, outside Aberdeen, in 1872 and had it rebuilt in 1881 in its present form.[10] Colonel Ogston also acquired Kildrummy Castle in Aberdeenshire. Nearby the old, ruined castle, the Colonel built New Kildrummy Castle in 1900.[11] The scale of these purchases and building projects – accounting for three stately homes in Aberdeenshire – conveys the affluence enjoyed by Ogston's near relations. It is possible the success of his uncle and cousins may have, in some measure, spurred Ogston's resolve to advance his own career – and shaped his desire for a country seat of his own, which he achieved by purchasing the estate of Glendavan, Deeside, in 1888.[12]

Ogston's father, Francis, gained a MA degree from Marischal College in 1821 and continued his studies at the University of Edinburgh, where he graduated MD in 1824. After a period on the continent, Francis Ogston established a medical practice in Aberdeen, where, in 1831, he was appointed City Police Surgeon. In 1839 Ogston was made Lecturer of Jurisprudence at Marischal College, and the lectureship was elevated to a professorial chair in 1857. Francis Ogston also served as Aberdeen's First Medical Officer of Health (1862–1881), was Dean of the Faculty of Medicine at Aberdeen, and served for two terms on the University Court as representative for the Senate. His *Lectures on Medical Jurisprudence* were published in 1878.[13]

In 1841 Francis married Amelia Cadenhead, daughter of Alexander Cadenhead, Aberdeen's Procurator Fiscal. Amelia's brother George took over his father's post after the latter's death in 1854. And George Cadenhead would remain in office until 1886, when he was appointed Aberdeen County Procurator Fiscal. The children of Francis and Amelia, therefore, grew up in an atmosphere of respectability emanating from the legal, medical, and manufacturing status of their maternal and paternal lines. Ogston's siblings, however, tended towards their father's academic and medical leanings. Ogston's sister Jane (1842–1923) married, in 1874, Reverend Henry Cowan, who became Professor of Church History at Aberdeen University in 1889. Ogston's brother Francis (1846–1917), meanwhile, became Professor of Forensic Medicine at Otago University, New Zealand. Finally, Ogston's sister Helen, born in 1848, married Archibald E. Malloch (1844–1919), a Canadian doctor. Having graduated BA from Queen's College, Kingston, in 1862, Malloch enrolled in the medical programme at Glasgow University under Regius Professor Joseph Lister. Like Ogston, Malloch was impressed with the results Lister had achieved in reducing post-operative infection through the use of carbolic acid to create an antiseptic surgical environment. Malloch graduated MB in 1867 and was appointed, through spring to autumn 1868, as Lister's house surgeon.[14]

On 19 April 1844, Ogston was born at Ogston's Court, 84 Broad Street, Aberdeen. His mother, Amelia, died in 1852, aged thirty-four. Ogston's only memory of her, he claimed, 'was her taking off her belt to thrash him'. In 1858 he attended The Gymnasium School in Old Aberdeen and from there matriculated, in 1859, to Marischal College, Aberdeen. Marischal College was founded in 1593, and would, during Ogston's time as a student, merge with King's College in 1860 to form Aberdeen University. This union was brought into effect by the Universities (Scotland) Act 1858, which stipulated medicine and law would be taught at Marischal College while theology and the arts would be the preserve of King's.[15] (Francis Ogston retained the Chair of Medical Jurisprudence following the unification of the Aberdeen colleges and would continue in that role until his retirement in 1883.) Ogston's undergraduate classes comprised Greek, Latin, mathematics, botany, and chemistry, alongside practical and comparative anatomy, physiology, chemistry, materia medica, medical jurisprudence, histology, microscopy, clinical surgery, and midwifery. Additionally, he attended Aberdeen Royal Infirmary and undertook a combined ten months' dispensary work.[16]

In the summer of 1863, following his fourth year of studies, Ogston embarked on a tour of the continent. An account of these days features in a series of biographical sketches called 'Scattered Recollections'.[17] In Prague, Ogston attended the city's Medical Society, which gathered in 'a small box, one of many in a public house'.[18] Ogston was joined on his travels by William Stokes (1838–1900) – son of the great Irish physician William Stokes (1804–78). Ogston recalls committing two – uncharacteristic – acts of callow levity in Stokes's company. When the pair visited Prague's Hradschin (or 'Castle District'), Stokes distracted a museum attendant while Ogston stole 'a handful' of hair from the tail of Albrecht von Wallenstein's stuffed horse.[19] The pair also visited Wartburg Castle, where Martin Luther translated the New Testament into German. Ogston and Stokes – for the price of 'a silver thaler' – were permitted by the attendant to cut out a small piece of Luther's table and even to remove a thumbnail 'of the manuscript of the Bible itself'.[20]

In Vienna, Ogston – where he was again accompanied by Stokes – enrolled as an extraordinary student at the university and attended lectures in anatomy, physiology, pathology, clinical medicine, and surgery. There, Ogston was also inducted as an extraordinary member of the student Verbindung (fraternity), which met on Sunday evenings in an inexpensive eating-house where the brotherhood engaged in singing, conversation, toasts, and smoking. Some among the students smoked cigarettes, and Ogston claimed that, having embraced this means of consuming tobacco, he 'was the first who smoked cigarettes in [Aberdeen] – much to the scandalising of the old-fashioned ladies and staid businessmen there, who considered me as given over to perdition for smoking a cigarette in Union Street in the day-time'.[21]

Ogston and Stokes visited Bohemia, Saxony, and Switzerland before they enrolled for the summer session at the University of Berlin. Here Ogston and Stokes studied under Rudolf Virchow (known as 'the father of modern pathology'), the pioneering ophthalmologist Albrecht von Gräfe, and Bernhard von Langenbeck, whom Ogston described as 'the leading surgeon in the world' in the 1860s.[22] Ogston and Stokes were, with the exception of 'three Swedes', the only foreign students in Berlin during the summer of 1864.[23] Their visit coincided with the German–Danish War, following which Denmark ceded the duchies of Schleswig, Holstein, and Saxe-Lauenburg to Prussia and Austria. Ogston notes that from the First Schleswig War (1848–52), Germany 'had commenced to form and work out into a system of policy their grandiose schemes for the domination first of Europe and next of the whole world'. Success in the German–Danish War and the subsequent Franco-Prussian war (1870–71) had the effect of further 'instilling into the nation the persuasion that they were a great and irresistible military power'.[24]

Ogston and Stokes also made walking tours in Thuringia and the Harz Mountains, following which Ogston studied for two months in Paris under Jules Germain François Maisonneuve and Charles Richet. He subsequently resumed studies at Aberdeen University, serving as Clinical Clerk to Dr William Williamson. In 1865 Ogston graduated MB, CM (with Highest Honours) and began general practice in

Aberdeen on 1 May, aged just twenty-one. He graduated MD the year following and opened an eye dispensary in Castle Street, which closed two years later.[25] Also in 1866, Ogston was made Assistant Professor of Medical Jurisprudence under his father, a role which earned him £25 per annum.[26]

As his professional career steadily advanced, Ogston had an increasingly large family to provide for. In 1867 he married Mary Jane Hargrave. Born in 1848, Mary Jane was the daughter of James Hargrave (1798–1865) from Hawick, Scotland, who superintended the Hudson's Bay Company's York Factory district in north-eastern Manitoba on the Hudson's Bay shore. In 1840 Hargrave married Letitia MacTavish (born in Edinburgh, 1813), the granddaughter of the chief of Clan Tavish and daughter of the Sheriff of Argyllshire. The young couple settled at York Factory and, in 1844, Hargrave was promoted to Chief Factor. A daughter, Mary Jane Hargrave, was born in 1848. Letitia died in 1854 and, in 1859, James married Margaret Alcock.[27] After James Hargrave's death in 1865, Mary Jane – known as Molly – was brought to Aberdeen by her stepmother. According to Janet Teissier du Cros – Ogston's granddaughter – this journey was undertaken to escape a suitor of Molly's whom Margaret deemed undesirable. Du Cros comments, '[Molly] had scarcely been a week in Aberdeen when she met my grandfather at a ball. He was already a brilliant young doctor and was so good-looking he came to be known as the "Adonis of the Profession".'[28]

This account of the couple's meeting is supported by another of Ogston's granddaughters, Molly Dickens. She adds: 'Molly and Alec were a handsome couple – handsome and explosive; she with her carefree manners, and he with his fiery temper. [. . .] Alec was tall, handsome and sandy-headed; Molly was petite, gay, with glorious auburn hair. They both laughed delightfully; they both fired up at a word.'[29] The marriage of Molly and Ogston produced a series of six near-contiguous pregnancies, resulting in four children – Mary (1868–1937), Francis (1869–1901), Flora (1872–1929), and Walter (1873–1957) – who survived into adulthood.[30] Dickens remarks, 'rather to his surprise I think, Alec saw his career threatened. Was his future to be snatched from him? Must he give up his ambitions and settle down to a hand-to-mouth existence, earning as a GP the living that would rapidly be required?' Ogston was determined otherwise. He sent his patients bills rather than follow the convention of accepting what remuneration they could manage or believed adequate. Determined to be more than a provincial doctor, Ogston read French and German newspapers and foreign scientific publications during his carriage rides between visits to patients.[31] To utilise even these imperfect interstices as an opportunity for study indicates the remarkable discipline and work ethic that would propel him to the heights of his profession.

Alexander was appointed, again alongside his father, as Joint Medical Officer for Health in Aberdeen in March 1868. The Assistant Professor and Joint Medical Officer roles indicate Francis Ogston was, in some measure, using his position to advance his son's career. When Ogston submitted his application for the post of Ophthalmic Surgeon at Aberdeen Royal Infirmary, however, he sourced twenty-nine supporting testimonials from medical professionals, including: William Pirrie (Professor of Surgery at Aberdeen University), John Struthers (Professor of Anatomy at Aberdeen), Robert Dyce (Professor of Midwifery at Aberdeen), George Ogilvie (Professor of Physiology at Aberdeen), George Dickie (Professor of Botany at Aberdeen), James Brazier (Professor of Chemistry at Aberdeen), Henry D. Littlejohn (Medical Officer of Health for Edinburgh), Robert Rattray (Resident Surgeon at Aberdeen Royal Infirmary), J. W. F. Smith (Physician at Aberdeen Royal Infirmary), William Farr (Superintendent of Statistics in the English Registrar-General's Office), and William Stokes (Surgeon to Richmond Surgical Hospital). The range and eminence of these references strongly argue Ogston, quite independently of his father's influence, had impressed senior colleagues as to his ability and promise. Robert Dyce, for example, noted Ogston 'enjoyed a more extensive and intimate knowledge of his profession than is generally experienced by men of his own years' and praised his 'zeal, assiduity and perseverance'.[32] Pirrie, meanwhile, declared Ogston was 'one of the most distinguished students of his years' and remarked, given his 'high abilities, his extended and liberal education, his natural turn and aptitude for surgery [. . .] I am quite sure [. . .] he would be a great acquisition to the staff of this Hospital'.[33]

An ambitious and promising young professional, Ogston was quite prepared to criticise ongoing practices at Aberdeen University and Royal Infirmary – institutions where he was attempting to establish his career. A memorandum to the University's Senate in October 1869 outlined Ogston's view that the medical curriculum lacked several important subjects, including diseases of the eye and ear, mental illness, dentistry, dermatology, and operative surgery. As a result, it was deemed students could electively enrol in supplementary classes taught by local professionals. Ogston's father secured permission for him to deliver summer classes in ophthalmology.[34] Being elective, however, these courses did not remedy the deficiencies Ogston identified. And it seems little had changed by 1877, when Ogston addressed the Aberdeen Medical Students'

Society. '[N]ot one-half of the students who graduate in our school,' he commented, attended the extramural ophthalmology class.[35] Not only were graduates unschooled in relevant branches of medicine, a disproportionate amount of time was devoted to preliminary sciences such as natural history, botany, and chemistry, which, though useful, did not 'constitute the chief knowledge required'.[36] Altogether, Ogston declared the Aberdeen medical student 'misled, overworked, and badly advised'.[37]

Ogston began his career at a revolutionary moment in the development of surgery. In the early Victorian era, the risks of operating were so great procedures were usually only carried out in unavoidable circumstances, such as amputations, tumours, and compound fractures. Aberdeen Royal Infirmary, for instance, admitted approximately 1,000 patients to its surgical wards every year between 1840 and 1844, 'but only some 80–120 operations, or two each week, were performed'.[38] Changes, however, were afoot. In the year of Ogston's birth – 1844 – Horace Wells, a dentist, witnessed the ability of laughing gas to numb pain at a demonstration given in Hartford, Connecticut. The following year, Wells unsuccessfully attempted to demonstrate nitrous oxide anaesthesia to an audience of Harvard medical students. In October 1846, however, William T. G. Morton, using an ether anaesthetic, carried out a successful operation on a male patient with a swelling under the right mandible. Learning of this new means of conducting painless surgery in December 1846, Robert Liston and William Squire carried out a successful leg amputation using ether anaesthesia at University College Hospital, London. Then, in 1847, Edinburgh physician James Young Simpson demonstrated the anaesthetic properties of chloroform, which became the standard anaesthetic in Scotland.[39]

During Ogston's time as a student at Aberdeen Royal Infirmary, 'it was still a matter of debate among surgeons whether chloroform [. . .] ought to be used in operations,' Ogston remarked, although chloroform was the sole anaesthetic 'ever thought of practically in Scotland [. . .] [u]sually it was not used'. Dr William Keith, a surgeon at Aberdeen Royal Infirmary, was 'the chief opponent of chloroform'. Keith – known as 'Old Danger' (presumably among Ogston's fellow students) – insisted 'his operations were more successful without it' and provided no pain relief, except a half glass of whisky for lithotomy patients. Remembering the sight of Keith cutting into the flesh of fully conscious patients, Ogston remarked: 'It seems [. . .] astonishing to look back upon it.'[40] By the time of his graduation, however, chloroform had become accepted and Ogston's duties, outside of those as Ophthalmic Surgeon, included serving as ARI's Anaesthetist.

Ogston recalled: 'We had all, without exception, been brought up to believe that suppuration was one of the necessary and inevitable "stages" of the healing of a wound.'[41] According to Ogston, James Young Simpson hypothesised infection was 'somehow connected with the little morsels of flesh that were snared off and remained in the wound when' ligatures were used to tie severed blood vessels. Accordingly, Simpson suggested acupressure – 'the employment of needles and wires [. . .] instead of thread, in securing the haemorrhage'.[42] When Ogston was appointed Ophthalmic Surgeon at ARI in 1868, the Infirmary 'was one of the few places acupressure was still in use'.[43] In fact, William Pirrie and William Keith had only recently published their *A Practical Treatise on Acupressure* (1867). Ogston was sceptical of Pirrie's claims for acupressure, however, including its propensity to prevent suppuration. The apparent success Pirrie had seen was, in fact, 'due to the zeal with which his trusted nurses in his wards wiped away every drop of pus immediately before' Pirrie made his rounds.[44] Ogston conducted experiments on arteries to assess the haemostatic properties of acupressure, torsion, and ligature. Gauging the ability of these methods to withstand internal pressure measured in inches of mercury, Ogston concluded 'ligature is still our securest means of arresting haemorrhage from wounded vessels'. These remarks were published in *The Lancet* in 1869. Ogston prefaced his conclusions by noting 'the observations of cases treated by acupressure in the Aberdeen Hospital would lead any impartial witness to believe it a most valuable plan in most cases'.[45] Despite this courtesy, Ogston's ambition, meticulous standards of inquiry, and willingness to challenge his superiors are demonstrated by this early publication.

Ogston's career continued to prosper. In July 1870, he took on the role of Junior Surgeon at ARI. In his application letter, Ogston, rather than 're-issue the Testimonials' he provided in application for the Ophthalmic Surgeonship, instead pointed to the eight academic publications he had authored in the interim.[46] In addition to his paper in *The Lancet*, by the time of submitting his application in 1870, Ogston had published on the excision of the calcaneum, the extra-capsular fracture of the femur, and removal of the posterior adhesion of the iris. Additionally, he produced four articles on general medical topics: weights of bodies and lungs of live- and still-born children, the function of the semi-circular canals of the inner ear, forms of sudden death, and spontaneous combustion. These eight publications had appeared in the *New York Medical Gazette*, the *British Medical Journal*, *British and Foreign Medical-Chirurgical Review*, and the *Medical Times and Gazette*.[47] Ogston, then, was establishing himself not only as an academic and surgeon, but as a researcher making original and wide-ranging contributions to medical science. Ogston was appointed Junior Surgeon, but, given there were few duties for him to undertake, he instead became Aurist to the Infirmary. Later that year, he resigned his Ophthalmic Surgeoncy and Practical Ophthalmology lectureship, retaining his Assistant Professor and Joint Officer for Medical Health posts.

Despite the miraculous advantages of anaesthesia, by modern standards, surgical hygiene was alarming. Describing the operating theatre at ARI, Ogston wrote: 'there was no appliance for washing the hands [. . .]. At the foot, or side, of the stained, coarse, old operation table was a wooden tray of sand, smelling of cats; on a shelf around lay the instruments, open for anyone to handle; and suture needles and acupressure pins were stuck ready in a jam-jar of rancid lard, which never seemed to be changed or cleaned'.[48] Surgeons' gowns, meanwhile, consisted of 'old, black coats covered with the dirt of years and encrusted with blood-stains, the dirtier the more venerated'.[49] Peter Jones summarises the state of affairs at this point in the mid nineteenth century:

> Chloroform was then the anaesthetic of choice, so surgery was no longer quite so 'butcherly' as it certainly was before 1846, but the range of operations performed was still limited – mainly amputations, the occasional mastectomy and cutting for bladder stones – while speed was still considered a virtue. The risks of post-operative wound sepsis were as real as ever, and conditions in the wards remained primitive.[50]

Ogston recalled ARI's 'wards, even the very corridors, stank with the mawkish, manna-like odour of suppuration'. Bloody gowns; instruments and hands seldom washed; cat droppings: it was 'small wonder' that '[n]ot a single wound healed without festering'.[51] In 1869, '102 ophthalmic operations were performed' at Aberdeen Royal Infirmary, including '16 for cataract'.[52] Ogston's remark that following 'every operation we used to await with trembling the dreaded third day, when sepsis set in', indicates his own procedures – and certainly those of his colleagues – regularly became septic.[53] This state of affairs is rendered more vivid by two examples offered by Ogston. One of the Infirmary's nurses 'preferred to die of a strangulated hernia', rather than submit to a seemingly fatal operation. Later, when Ogston assumed command of the operation ward, he tore down and burned a sign declaring 'PREPARE TO MEET THY GOD' – much to the consternation of his colleagues.[54] It would take Joseph Lister's pioneering work in introducing antiseptic methods to alleviate this situation.

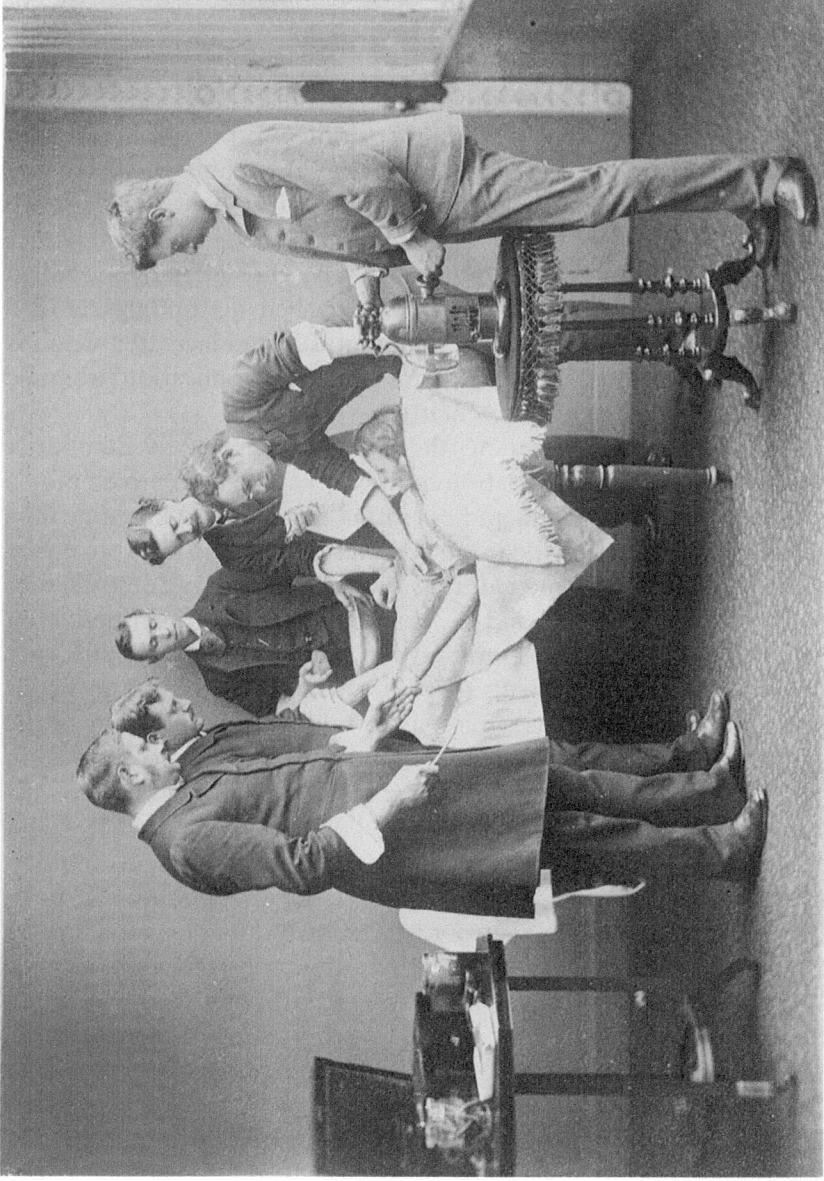

Figure 1.1 Carbolic acid spray being used at one of Ogston's surgical procedures at Aberdeen Royal Infirmary. © Aberdeen City Council (Archives, Gallery & Museums Collection).

Born in Upton, Essex, Lister (1827–1912) graduated MB from University College, London (1852) and thereafter became Assistant Surgeon at Edinburgh Royal Infirmary (in 1856) and Regius Professor of Surgery at Glasgow University (in 1860). In 1864 Lister was alerted to the work of Louis Pasteur, who, as Lister summarises, had established: 'the septic property of the atmosphere depended not on the oxygen or any gaseous constituent, but on minute organisms suspended in it'. Lister, therefore, reasoned that 'decomposition [. . .] might be avoided [. . .] by applying as a dressing some material capable of destroying the life of the floating particles'.[55] Lister found carbolic acid applied with swabs or dressings could successfully disinfect contaminated tissues in lacerated wounds and compound fractures. He successfully applied these principles to eleven patients and published his findings in *The Lancet* during 1867. Lister reasoned if 'contused and lacerated wounds heal thus kindly under antiseptic treatment [. . .] its application to simple incised wounds must be merely a matter of detail'.[56]

In 1869 Lister was elected to the Chair of Surgery at Edinburgh University. Following his relocation to the capital, he began to employ a spray that covered the wound and operating environment in 'a fine mist of carbolic acid'. This provided some reassurance against the threat that '[a] floating germ might enter during the operation into some cellular interstice among the tissues and [. . .] spread putrefactive fermentation through the wound'.[57] On lecture days in the Surgical Hospital, Lister would discuss a series of patients that were brought onto the stage by his dressers. If the required operation was to be performed in front of the class, Lister employed three assistants – one for instruments (soaked in 1-in-40 carbolic lotion), one for carbolic spray, and another to administer the chloroform anaesthetic (via a towel on the patient's face).

Ogston was thrilled, though incredulous, to learn Lister had apparently established a means of avoiding sepsis and suppuration in postoperative wounds, and he visited Lister shortly after he took up his position at Edinburgh University. '[W]ithout any introduction, I called on Lister at his own house and was received, though unknown, with all the sweetness and gracious courtesy which was a part of his nature'. Ogston also spoke to Lister's former assistant Hector Cameron, who was still working at Glasgow using Listerian methods. There, Ogston was shown a knee joint that had healed perfectly after operation. He was introduced to further examples from the wards, but Ogston was convinced within '[f]ive minutes [. . .] of the truth of the marvellous discovery'. 'I felt inclined to sit down, cover my face in my hands, and think out what the great revelation implied in the future.'[58]

Thereafter, in Aberdeen Ogston 'introduced the methods of Lister into the wards'; however, implementing these innovations 'was not without difficulty'. The Infirmary authorities baulked against the increased expense, while 'the older members of the Staff were indifferent, if not actively hostile'.[59] Ogston, however, became a fervent advocator of Listerian methods, and his enthusiasm for antiseptic practice was so profound it provoked facetious remarks from students. In an Aberdeen University graduation programme, the following ditty appeared:

The spray, the spray, the antiseptic spray,
A.O. would shower it morning, night and day.
For every sort of scratch,
Where others would attach
A sticking-plaster patch,
He gave the spray.[60]

Another student, meanwhile, noted 'Dr Ogston is nothing if he is not an Antiseptician. Antiseptics for this! Antiseptics for that! Antiseptics for everything! [. . .] All the superstitions of the past are to be forgotten in the Antiseptic Gospel of the present.'[61] While Ogston correctly realised the revelatory importance of Lister's innovations, he continued to use the spray until 1890–91, when it had ceased to be employed elsewhere. By 1887 Lister conceded the spray was not necessary to antiseptic practice.[62] Moreover, antiseptic measures were overtaken by aseptic ones. Rather than merely destroying bacteria present in the operating environment, aseptic measures eliminated the presence of bacteria altogether. In 1885 Ernst von Bergmann used steam to sterilise surgical instruments and extended this measure to dressings and gowns. It would be one of Ogston's most impressive pupils, Henry Gray (1870–1938), who introduced these measures to Aberdeen.[63]

In early 1872, Ogston was appointed Medical Officer to the Aberdeen Smallpox Hospital, and in February that year there appeared a note in the *British Medical Journal* stating Ogston was one of five Scottish doctors who had 'undertaken to give their assistance in aiding the extension of the Association in their respective localities'.[64] In March, the *BMJ* recorded the establishment of the Aberdeen, Banff, and Kincardineshire branch of the British Medical Association – the Association's first Scottish division – and commended the 'efforts of Dr. Alexander Ogston, Aberdeen, who has spared no pains to contribute to the establishment of the Branch, and who will act as its honorary secretary'.[65] Again, and in another guise, Ogston was making a name for himself within the profession.

Ogston resigned his position as Medical Officer for Aberdeen in August and in November was elected a Fellow of the Medical Society of London. He further emerged from his father's shadow when, in September 1873, he resigned his Assistant Professorship in Medical Jurisprudence. His younger brother, Frank, succeeded him and remained in the Assistant Professorship role until the retirement of Francis Ogston in 1883. Frank was unsuccessful in following his father to the Chair of Jurisprudence, however, which accounts for his emigration to New Zealand.[66] In October 1873, Ogston became an Examiner in Medicine at Aberdeen University. Then, in June 1874, he was made a full surgeon at ARI.

The strain of six pregnancies in as many years took its toll on Molly, who committed suicide in 1873 – following the birth of her son Walter in November. It is possible she was suffering from what would now be diagnosed as postnatal depression. Du Cros notes Molly 'would sit list-less in her room with the blinds drawn' and that, though Ogston would open them before he left for the day, 'as soon as he was gone she would draw them down again'. Molly, perhaps feeling neglected, 'committed suicide with the help of some drug from her husband's laboratory'.[67] According to Molly Dickens, her grandmother's suicide occurred three days after Christmas Day 1873. She was twenty-five years old.[68]

For du Cros, Ogston's busy working life meant he 'failed to take seri-ously the loss of spirits and the depression that wore [Molly] down'. Additionally, she claims, he, 'like so many Scots, he had begun on occasion to drink too much'.[69] The claim of Ogston's drinking appears incongruous given the aura of rectitude he exuded in later life. Dickens corroborates these claims, but notes, however, after Molly's death 'there was no spectacular reform (Alec was not like that)'.[70] Rather, he for-swore immodest consumption of alcohol, and there is little evidence to suggest he was anything but an infrequent drinker thereafter. Ogston never fully recovered from his grief at Molly's suicide, in which he must have felt to some degree implicated. Although he later married Isabella Margaret Matthews (1848–1913), upon his death Ogston was buried alongside his first wife in the Ogston family plot in St Clement's church-yard, Aberdeen. According to his granddaughter Mary, Ogston kept a portrait of Molly in his study throughout his life, despite the request of his second wife that it be removed.[71]

Following Molly's death, Ogston's sister Jane 'stepped into the breach' and assisted with the domestic situation, alongside the family nurse, Isy.[72] Ogston's son Walter also notes the family had a cook, Katherine Eddy.[73] The family home at 252 Union Street was an imposing build-ing. It possessed a double basement (including a wine cellar) where the servants worked. A stone staircase 'muffled in turkey [Turkish] carpet' led to the first floor where there was a drawing room containing Molly's piano – around which she taught the children hymns such as 'Once in Royal David's City'. Nearby was the parental bedroom, which Dickens remembers as 'look[ing] down on a grave-like garden, dug deep below the granite cliffs'.[74] She remarks: 'There were no bathrooms anywhere, none anywhere.' On the second floor, however, in addition to a guest room and sewing room, there was a cold-water tap. From this, the domestic staff had to 'supply all the bedrooms with hot baths, and all the hand-basins in all the bedrooms with both hot and cold water'. From the second floor, a further staircase led to the nursery. Dickens remembers this being a more light-hearted area, where 'all was joy' – where there could be found 'an aunt rattling on the piano, an aunt making toast before the fire on a long-handled toasting-fork' and even 'an aunt teasing an uncle'.[75]

Walter's memoirs provide a glimpse into a respectable, though imperfect, domestic situation. Ogston, a widower at twenty-nine, worked multiple demanding jobs to support four children and retain a domestic staff. Unsurprisingly, he had limited discretionary time. Walter's account of his early childhood is dominated by the nurses – first Isy, then Sarah – that raised him, and latterly by the family governess, Miss Maxwell. 'I saw little of my father. He was a very busy man and had little time to spare.' Each night, the Ogston children spent half an hour at the dining room table where they 'looked at books or played quiet games'.[76] According to Walter, Ogston – whose study was next to the dining room – sometimes joined the children and sat in an armchair in the room.[77] He 'seldom spoke to us', Walter recalled, but instead read or played patience. Walter describes a solemn bedtime ritual where Ogston 'held out his closed hand' to be kissed by each child, replying to each a 'Goodnight' in a 'loud voice'.[78]

Walter does mention lighter moments, adding there was 'some romping' at these evening gatherings. Chestnuts were roasted on the fire and the siblings stuck 'postage stamps into little books which [Father] had given us'.[79] On Sundays, Ogston taught his children paraphrases and awarded a 'pink sweetie' to the most adept. As a conclusion to these sessions, each child was given a piece of flavoured rock from the 'rock box' that Walter carried through from his father's study.[80] Undemonstrative by nature and burdened with a taxing professional life, Ogston, as a father, emerges as a distant, revered figure. What occupied his thoughts during these scant moments of recreation? Perhaps, given the standards of the time, there was more domestic felicity than a modern sensibility detects in Walter's depiction of an apparently strained family life. But how far was Ogston gratified by the companionship of his children? Did he feel his steadily burgeoning career rewarded the exertion and sacrifice that went into it? Or did Molly's absence from these evening gatherings raise the possibility he had staked too much on professional advancement? In any case, his labours had forged a career that was gathering momentum – one that would take him to the forefront of contemporary medicine.

2

1874–1882

My delight may be conceived when there were revealed to me beautiful
tangles, tufts, and chains of round organisms in great numbers.[1]

OGSTON WAS APPOINTED FULL SURGEON at Aberdeen Royal Infirmary in
June 1874, and in 1877, four years after Molly's death, he married Isabella Matthews (1848–1913), daughter of the prominent Scottish
architect James Matthews (1819–98), who designed Aberdeen's Grammar
School and Art Gallery. James Matthews – who later served as Lord
Provost of Aberdeen (1883–86) – also designed several Scots baronial
stately homes, including Ardoe House and, in 1888, Ogston's country
home of Glendavan.[2] By the accounts of Ogston's children and grandchildren, his marriage to Isabella – or 'Bella' – introduced a degree of strain
into the family home. Molly Dickens – Ogston's granddaughter through
his first daughter, Mary – remarks: 'Even her own delightful children
did not fill the blank in Bella's heart – never a warm or capacious one'.
Dickens suggests Bella was prone to melancholy, writing she 'spent so
much of her life on the sofa recovering from her down-lyings' caused by
the 'mortal disease from which Bella suffered – and made others suffer'.[3]
Regardless, the Ogston family continued to grow, with seven children
resulting from his second marriage. The expanding family necessitated a
further storey being added to the house at 252 Union Street.[4]

Janet Teissier du Cros – also Ogston's granddaughter through Mary –
describes Bella as 'a woman we none of us loved', who ingratiated herself into Ogston's life largely through appeals that his children required
a mother figure.[5] Walter likewise recalled, albeit in more sympathetic
terms, Ogston's second marriage as a moment of profound alteration:
'Life for me was revolutionised and in relation to him was entirely
changed – though no one was to blame for this – after his marriage to

my stepmother. As the care of us four children [those from Ogston's first marriage] naturally fell to our stepmother and nurse, my main affection was diverted to the former and Father seemed to recede into the background.' Expressing the situation charitably, Walter observed: 'I will only say that the children of the first family, when they had become independent and had left home, had a real affection for their stepmother.'[6]

Walter was equally understanding of his father's scant presence in his children's lives: 'We did not see a good deal of him, for he was a young man, devoted to his profession, and had little time to give to his family.'[7] Still in his early thirties, Ogston evidently set his sights above being a provincial surgeon. Accordingly, alongside his duties at ARI, Ogston applied himself to a variety of investigative and research work. During this period, he published papers on the origin of cancer in the epithelium (1873),[8] collated known knowledge on congenital malformation of the lower jaw (1874),[9] and argued for the existence of oblique fractures in the head of the humerus (1876).[10]

Ogston was working during an exciting time for surgery, when Lister's aseptic innovations opened possibilities for advances in elective surgery (as opposed to operating only in the last resort). One of Ogston's most important surgical innovations was a refinement of the operation for *genu valgum*, a skeletal deformity which may be caused by vitamin D deficiency. In individuals with *genu valgum*, stress on weakened leg bones results in a postural malformation where an individual's feet are far apart when they stand with both knees together. Rather than surgically removing the majority of the femur's articular surface, as Thomas Annandale's 1875 operation called for, Ogston pioneered a process which allowed the detachment of an elongated inner (medial) condyle of the femur to slide upwards, thereby bringing it more into line with the outer (lateral) condyle.[11] Ogston successfully carried out this procedure in summer 1876. The following year, he delivered a paper on the operative treatment of *genu valgum* at the annual Congress of the German Surgeons in Berlin and published an account of his operation in the *Edinburgh Medical Journal*.[12] Ogston's procedure paved the way for William Macewen's improved treatment for *genu valgum*, established in 1877, whereby the deformity was treated by osteotomy of the femur, rather than altering the knee's articular surface.[13]

Ogston made bone structures a research speciality during this phase of his career. In addition to work on *genu valgum*, he carried out considerable investigations into the nature of articular cartilage. In an 1875 paper published in the *Journal of Anatomy and Physiology*, Ogston demonstrated that, instead of being 'mere cushions to diminish shock and render motion smooth', articular cartilage was 'really a growing tissue', one 'as valuable

Figure 2.1 Photograph of Ogston by R. M. Morgan Ltd. Wellcome Collection. Public Domain.

and necessary in forming and maintaining the structure and shape of bone as periosteum [a membrane covering the outer surface of bones] is always admitted to be'.[14] Developing these findings in an 1878 article, Ogston demonstrated that the 'main beams or trabeculae' of articular cartilage are 'placed at right angles to the bone surface, while periosteal bone is characterised either by its main trabeculae being parallel to the bone surface or by there being no indication of any special direction observable in them at all'.[15] Peter G. Bullough describes these findings as 'fundamental to the understanding of joint anatomy'.[16]

Also in 1878, Ogston published an account of his treatment for club foot. Disappointment in tackling this condition emanated, he claimed, from 'the pernicious doctrine, that ordinary congenital club foot depends on paralysis or contraction of certain muscles and fasciae'.[17] Infants born with this deformity, he pointed out, experienced 'neither paralysis nor contraction'. Rather, all constituents of the limb – skin, muscles, tendons, fasciae, and bones – contribute to the malformation. Ogston proposed, therefore, that *every structure in the limb, without any exception, contributes its share towards keeping up the malformation, and has to receive its share of the treatment*.[18] The procedure he implemented used the existing method of resetting the limb in plaster of Paris. However, Ogston suggested a method whereby, after administering chloroform anaesthetic to the patient, the operator used their hands to manually manipulate the limb 'until the resistance of the foot being by degrees overcome, it unfolds from its perverse position' and is then set in plaster.[19] Ogston did not believe corrective orthopaedic apparatus to be any more effective than his method, and recommended – given the former treatment was beyond the means of many patients and medical institutions – his procedure as the most effective for treating club foot among poorer patients.

Ogston's greatest discovery, however, was obtained by pursuing the logical next step indicated by Lister's work. Now surgery had largely been freed of the scourge of fatal post-operative infection, Ogston was compelled to ask what, precisely, 'is the cause of acute suppuration, of acute inflammation, and of blood poisoning after wounds and operations?' He reasoned:

> Often I meditated on the subject and became the more convinced that there was a single cause, and that the cause was some special germ. But it was some time before it was possible for me to verify this conviction.[20]

In his summary of bacteriological science at this historical point in the Victorian era, George Smith notes 'for the most part the concept of bacterial disease in man was most confused, indeed chaotic'.[21] Antonie van

Leeuwenhoek (1632–1723) – a Dutch haberdasher, draper, and lens maker –
was the first to observe bacteria, but it was 'only after 1870', once 'the
association between specific microorganisms and infectious diseases' had
been postulated, 'that the study of bacteria was recognized as important
for medicine and public health'. Pasteur's work on fermentation 'demon-
strat[ed] that different types of microbial organisms were responsible for
different types of fermentation'.[22] However, bacteriological research had
by no means arrived at the germ theory of disease, whereby particular
microorganisms are identified as being the cause of specific diseases.

When Ogston began his research, heterogenesis – the 'spontaneous
generation of microorganisms from organic matter' – was still defended
by some, such as Charlton Bastian, author of *The Beginnings of Life*
(1872).[23] Additionally, the theory of pleomorphism – that only one form
of bacteria existed and this changed depending on its environment –
was maintained by Carl Nägeli and others.[24] Meanwhile, as Ogston
described it, 'Lister himself was vague on the subject' of suppuration:

> He attributed it generally to germs, but held also that there were other
> causes, such as tension, for the latter of which he wrote strongly in favour of
> the drainage of wounds and wound cavities. Now it was evident that in this
> last opinion he was wrong, for a great number of instances were known in
> which vast tension, such as in ovarian and other tumours, existed without
> any suppuration ensuing.[25]

Lister's views, however, were 'not shared by many other workers' in the
field. And connections were being made between micrococci and erysipelas, puerperal peritonitis, and endocarditis.[26] Means of securing evidence
for such causal links was offered by the work of Heinrich Hermann
Robert Koch (1843–1910), whose *The Aetiology of Traumatic Infective
Diseases* (1878) demonstrated small animals could be infected with particular diseases via injection of putrid fluids. Koch pointed the way forward to Ogston by calling for a 'series of similar experiments on animals
with materials obtained from persons suffering from, or who had died
of, traumatic diseases (septicaemia, pyaemia, progressive suppuration,
gangrene and erysipelas) and – what seems to me to be the most important, to look for microorganisms in the human body'.[27]

An opportunity to test his theory that a 'special germ' was responsible for acute inflammation was presented to Ogston when he attended
'a Mr. James Davidson [. . .] a young man suffering from an extensive
suppurating phlegmon of the leg':

> Procuring a clean file, I evacuated into it the matter from the phlegmon
> through the unbroken skin, preceded home with it, and placed a little of the

pus under an ordinary student's microscope fitted with a quarter inch objective. My delight may be conceived when there were revealed to me beautiful tangles, tufts, and chains of round organisms in great numbers, which stood out clear and distinct among the pus cells and debris, all stained with the aniline solution I had employed to render them more distinct.[28]

This finding 'appeared to indicate the solution to a great puzzle'.[29] However, the logical next step was to determine conclusively the microorganisms in the sample obtained from Davidson were not present coincidentally but did, in fact, represent the precise microorganism responsible for acute suppuration.

Calling upon his medical contacts, Ogston began examining examples of abscesses supplied by his colleagues. Contributors included his brother, Frank, and doctors with practices in Thurso, Stonehaven, Ballater, and Peterhead.[30] Using a Zeiss microscope obtained with a British Medical Association grant, Ogston examined 'a series of cases – between eighty and a hundred – and found ample corroboration of the existence of the micrococci in all instances of acute suppuration'.[31] Having established the presence of micrococci in suppurative wounds, Ogston then sought to demonstrate 'whether acute inflammation and suppuration could be produced by pure cultivations of micrococcus germs'.[32] He built a laboratory in the garden behind his house at 252 Union Street and – using newly laid eggs incubated at the temperature of the human body – cultivated pure pus microorganisms, which were transferred subcutaneously to rats and white mice, thereby establishing the isolated micrococci were, in fact, responsible for suppuration.

Ogston initially shared these findings in his April 1880 paper, 'Ueber Abscesse', which he delivered at the Ninth Congress of the German Chirurgical Society in Berlin. Ogston provided an outline of his methodology and addressed the claims of Watson Cheyne (Lister's assistant surgeon at King's College, London) 'that micrococci are innocuous organisms'. Cheyne believed because micrococci can survive 'in diluted carbolic solutions[,] they frequently penetrate from outside into wounds treated' by an overly weak solution of carbolic acid.[33] Ogston's cultivation of pure pus micrococci and subsequent injection of guinea pigs convinced him, however, that '[m]icrococci are the most frequent cause of the formation of acute abscesses'.[34] Furthermore, Ogston shared two additional findings he would later develop. Firstly, he noted the existence of 'different kinds of micrococci': some 'in the form of chains' and other 'spherical bacteria in groups' that 'looked like bunches of grapes' (Ogston would later name this *Staphylococcus*).[35] Additionally, Ogston postulated a link between 'micrococci and blood poisoning'.[36]

His findings were lauded by the audience, who afforded him the honour of being made, at the age of thirty-six, a Fellow of the German Surgical Society. That year, his star continuing to rise, Ogston also became Senior Surgeon at Aberdeen Royal Infirmary.

The period leading up to 1882 was Ogston's most active as a researcher. From his academic publications, traces of the man emerge: a clear, powerful mind; an enterprising individual, surely, to research and publish these medical advances in addition to his other duties; nor, it is evident, was he afraid to challenge established opinion. Ogston, however, also left behind a series of personal travel journals. And, after publishing his March 1881 'Report on Micro-Organisms in Surgical Diseases' – an updated version of the Berlin address – in the *British Medical Journal*, he departed for a trip to Greece on 7 April.

Although he did not always apparently write or preserve a diary of his foreign excursions, the journal Ogston kept on this trip (titled 'A Run to Greece') reveals different sides of his character. He questioned the decision to embark on this journey:

> What on earth am I going for? As far as I believe, in the hope that sitting on the Acropolis of Athens (it is sure to be a wet seat) and probably smoking there, while regarding the remains of a great past, may open my mind and bring me nearer to understanding or being content to fail to understand the great question which has never been answered – what is man and what will he afterwards be?[37]

'A Run to Greece' shows aspects of Ogston's inner life that may have seldom been perceived by his contemporaries. Here, the redoubtable new Senior Surgeon cheerfully envisaged himself smoking, with a wet behind, and *failing* to understand his objective. Given the sacrifices Ogston made to forge a distinguished career and raise a large family, it is comforting to learn the occasional severity he projected to the outward world – even to his own children – was leavened by an element of self-deprecating whimsy. Furthermore, Ogston also liked to intersperse his journals with ink and pencil drawings – a habit which chimes with his observational bent of mind and reveals a minor artistic ability.

Enjoying the novel luxury of his surroundings on the south-bound train from Aberdeen to London, Ogston remarked:

> He who travels in a first-class carriage is for the time being an aristocrat, beholding, as from a private box, the theatre of the world. To him the porter cometh, the guard toucheth his hat, while the democracy around him, starting back with awe if it open by mistake his door and behold the grandeur of his feet upon the cushions, is herded about like a sheep, and has its doors slammed upon it.[38]

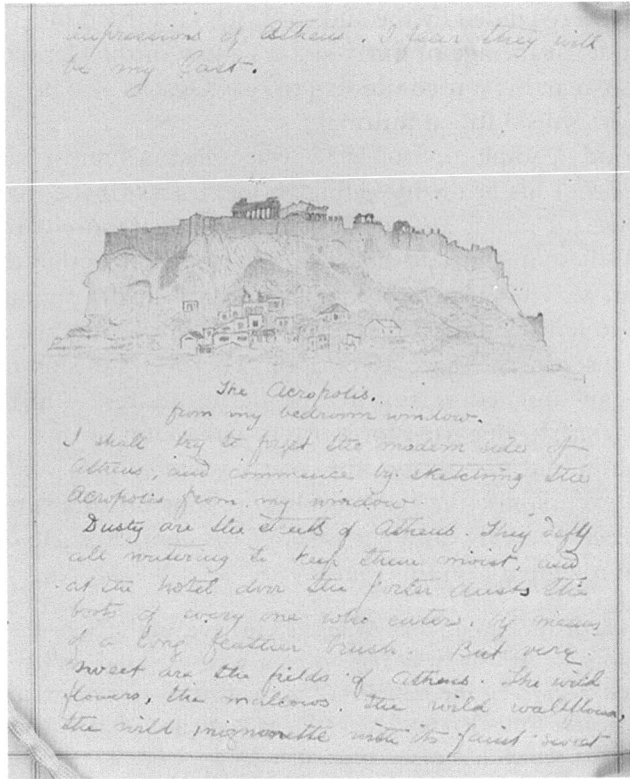

Figure 2.2 Sketch by Ogston of the Athenian Acropolis. MS-3850-1-1-00079 in the University of Aberdeen Museums and Special Collections, licensed under CC By 4.0.

Reflecting on the relative opulence of his carriage, Ogston recalled a Latin phrase from Virgil's *Georgics* (*c*.29 BC) 'Felix, qui potuit rerum cognoscere causas' [Happy he who was able to know the causes of things] – no doubt with the thought he had earned his moment of luxury.[39] Suggesting a retentive memory and significant awareness of literature, Ogston frequently inserted such quotations in his diaries. The journey to Greece includes references to John Bunyan's *The Pilgrim's Progress* (1678–94), Virgil's *Aeneid* (*c*.29–26 BC), and the poetry of Walter Scott. Ogston crossed the English Channel and continued by rail across the Alps into the Italian port of Brindisi, where he boarded a seemingly deserted steamship. Lodged in 'a black hole of a cabin' and apparently concerned for his safety, Ogston slept 'with [a] loaded revolver under [his] head'.[40] The next morning, cheered by a good breakfast and the sight of porpoises playing around the ship, he was relaxed enough to remark that the – still retained – revolver 'sleeps quietly in my coattail pocket'.[41]

Arriving in Athens on Saturday the 16th, Ogston made his way to the Acropolis. His recording of this moment, written directly in pen in his journal, demonstrates an admirable ability to evoke a scene:

> It is deserted at this hour of the morning, and from its highest point, seated on a marble chair in the temple surrounded by the views of a great past, broken columns, buildings overthrown, a giant quarry of marble fragments, the plains & hills of Greece spread out beneath and around me, the sounds of the city beneath, & the tinkling of the bells from its curious tiled churches, deadened by the distance as they are wafted up to me in the mild morning under the blue sky: I feel that I am richly repaid for my journey.[42]

While enamoured with ancient Athens, Ogston found the modern city filthy and depressing – sentiments he emphasised with unflattering comparisons to Aberdeen, claiming 'the worst quarter of the Gallowgate is finer than [Athens's] finest street'.[43]

On 20 April, Ogston left for Marathon with 'an old most villainous looking Greek' guide, and made subsequent excursions to Aegina, Zante, and Corfu.[44] From today's perspective, the conditions endured by travellers at this time were primitive in the extreme, yet Ogston revelled in the adventure. On-ship to Corfu, he recalled 'a pretty rough sea, & rolling severely'. Later that day, he noted with approval:

> Out of seven who turned up at dinner, four were Englishmen: and of the five who stood it out, three were Anglo-Saxons, well done, John Bull! I told them a number of gentlemen's stories, and we all laughed until we forgot to be sick.

Evidently a good sailor, Ogston was preoccupied with another misfortune: 'Calamities do not come singly, my tobacco has run done! This is the worst of miseries, and they keep none on board [. . .] If only I had a cigar!'[45] It would not, perhaps, surprise readers of his medical publications to learn Ogston was constitutionally suited to withstand challenging circumstances. However, the image of the occasionally gregarious, cigar-smoking, revolver-carrying figure which surfaces in 'A Run to Greece' allows for a more nuanced estimation of Ogston's personality to emerge.

Back in Britain, Ogston's conclusions regarding microorganisms were met with scepticism. His colleagues at the Aberdeen branch of the British Medical Association disputed his postulation of a 'causal connection between micro-organisms and inflammatory disease'.[46] Meanwhile, after publishing his 'Report on Micro-Organisms in Surgical Diseases', the *BMJ* refused to accept further papers by Ogston. According to Ogston,

BMJ editor, Ernest Hart – denigrating the apparently parochial origins of Ogston's theory – 'asked scoffingly: "Can any good thing come out of Aberdeen?"'[47] 'Very different,' claimed Ogston, 'was the reception which Lister gave to the discovery'.[48]

In the London home of Dr Matthews Duncan, Ogston demonstrated his findings to Lister, whom Ogston recalls having 'got to know [. . .] pretty well by this time'. Lister – who had relocated to London in 1877 to take up the Chair of Systematic Surgery at King's College Hospital – apparently 'expressed himself satisfied', although his assistant – Cheyne – was 'very sceptical'.[49] Ogston, however, made these particular remarks in 1920, when he was in his mid seventies. Retrospectively addressing the initial reception of his research, Ogston preferred to omit the reality that Lister, in fact, openly criticised his findings at the 1881 International Medical Congress in London. Addressing the Pathological Section of the Congress (which included Koch) on 5 August, Lister acknowledged 'the importance of the relations of micro-organisms to diseased processes in wounds', but, nevertheless, introduced his presentation as 'a needed note of warning against a tendency to exaggeration in this direction'. While '[a]cute inflammation is certainly very often caused by [. . .] minute organisms', Lister also claimed 'inflammation is often caused otherwise, viz, through the nervous system'.[50] In light of this contention, he advocated the use of counter-irritation in treating inflammation. According to Lister, '[w]hen two parts are nervously in sympathy with each other, if we excite a great action in the nerves of one we may distract action from the nerves of the other'. This method could be used, he maintained, to cure inflammation, thereby proving 'the inflammation so cured was maintained by an abnormal action of the nerves'.[51]

Turning directly to Ogston's research, Lister noted Ogston had 'not found micrococci present in any' chronic abscesses. And, although these had been found by Ogston in acute abscesses, Lister contended, 'Dr. Ogston leaves us entirely without any explanation as to the origin of the infection in the part in which the abscess occurs.' Lister supported this contention by observing that, while micrococci could be supposed to cause 'suppuration' and 'the inflammation which precedes it', inflammation could 'be induced by some altogether accidental circumstance'. Indeed, Lister went so far as to claim 'micrococci are, so to speak, a mere accident of these acute abscesses, and that their introduction depends upon the system being disordered'.[52]

Ogston responded with a lengthy paper entitled 'Micrococcus Poisoning', published in the July 1882 edition of the *Journal of Anatomy and Physiology*. Ogston summarised his findings to date: he had shown 'acute inflammation is capable of being produced by micrococci', demonstrated micrococci 'are the cause of acute suppurative inflammations', and established a 'close relationship between' micrococci and blood poisoning.[53] Ogston claimed he was 'right loath to disagree' with Lister but confident the elder surgeon would 'think no evil of those who contradict him in the search after truth'.[54] There was, Ogston insisted, 'no satisfactory evidence anywhere existing that would lead us to believe that counter-irritation has any influence on inflammation'.[55] He continued:

> the amount of true scientific evidence as to the causal connection between micro-organisms and acute inflammation is very great, while similar evidence as to the efficacy of counterirritation does not exist, and, therefore, I cannot abandon the position I have taken up concerning the dependence of acute inflammation on micro-organisms.[56]

Having dealt Lister this polite but comprehensive retort, Ogston developed his work on blood poisoning. For Ogston, the term 'blood poisoning' misleadingly implied 'the micro-organisms that cause diseases to *grow and multiply in the blood*'.[57] Ogston, however, questioned the notion blood itself is 'the chief site of the growth of the organisms'.[58] Rather, he proposed viewing blood poisoning not as 'a poisoning of the blood *per se*, but as a disease existing in the tissues, from which the blood is but secondarily infected'.[59] He concluded, therefore, that 'septicaemia (bacteria in the blood) and pyaemia (abscesses scattered round the body) were due to a localised source of micrococci infection which subsequently spread'.[60] And, for the first time, Ogston clearly differentiated between *Streptococcus* (which he linked with erysipelatous disease located in the lymphatics) and another micrococcus Ogston named *Staphylococcus* (which he connected with 'suppurative inflammation expending itself on the tissues').[61] The name *Staphylococcus* was suggested to Ogston by William D. Geddes – Professor of Greek at Aberdeen University (later Principal of the University) – as the Greek word *staphyle*, meaning 'bunch of grapes', formed an apt description. Ogston, furthermore, argued disease produced by micrococci would depend on the virulence of the cocci and the varying susceptibility of individuals.

Ogston's contributions to understanding the etiology of inflammation and septicaemia were contemporaneous with other important advances in the field of bacteriology. Koch described the bacteria responsible for tuberculosis in 1882, and other bacterial sources of disease were determined, such as diphtheria (Loeffler, 1884), cholera (Koch, 1884), tetanus (Kitasato, 1889), dysentery (Shiga, 1898), and pneumonia (Sternberg and Pasteur, 1881).[62] Today, the germ theory of infection is taken for granted, even to the extent of being widely understood by the general public. When evaluating the significance of his achievements, it is worth bearing in mind Ogston, at a relatively young age, advanced these ideas in the face of considerable criticism from Lister, Cheyne, Hart, and senior Aberdeen colleagues.

Despite Lister's resistance to his research, Ogston paid him a handsome compliment, writing to Lister in 1883 to say: 'You have changed surgery, especially operative surgery, from being a hazardous lottery into a safe and soundly based science.'[63] It was Ogston, however, who provided the decisive link between Lister's instigation of antiseptic surgery and an understanding of suppurative processes – one which Lister did not personally comprehend. According to Linder, while Lister's research generated 'world-wide dispute about the significance of micrococci in causing wound infection', it was Ogston's investigations that 'played a decisive role in settling the controversy'.[64] Lyell, similarly, remarks: 'The miracle that Lister started empirically, Ogston finished scientifically, and the memory of these giants should be linked, like some great binary star that has illuminated a very dark corner in the sufferings of mankind.'[65] In addition to his work on *genu valgum* and articular cartilage, Ogston established the link between acute inflammation and suppuration and microorganisms, discovered (and named) *Staphylococcus*, and correctly linked localised microorganism infections with blood poisoning. Undoubtedly, his insights merit a higher degree of general renown than they have received. Recalling the reception of his pioneering research, Ogston, however, noted, 'rebuffs did not really give me much concern. The men who mattered knew. The others were of little consequence.'[66] Ogston was aware his research had only partially illuminated the emerging field of bacteriological studies. He declined to continue his investigations, however, referring to 'the limited time and opportunities I possessed, in the midst of a big surgical practice'. Rather, he left 'to others, more fortunately situated, the pursuit of further enquiries'.[67] In 1882 he succeeded William Pirrie as Regius Professor of Surgery at Aberdeen University.

Ogston demonstrated 'cocci were the etiological agents of disease conditions' and 'was the first to recognise two forms of cocci' (*Streptococcus* and *Staphylococcus*).[68] In 1884 Julius Rosenbach further distinguished between staphylococcal colonies of different colour. Those that were white he named *Staphylococcus albus*, and those that were yellow in appearance he called *Staphylococcus aureus*.[69] *Staphylococcus aureus* occurs naturally in the skin and nostrils and is carried by approximately one-third of the human population. Upon entering the bloodstream, however, it can cause a variety of disorders, from skin infections to fatal septicaemia. Initially, all *Staphylococcus aureus* strains were susceptible to penicillin. However, 'in the early 1960s, the first strains of methicillin-resistant *Staphylococcus aureus* (MRSA) emerged'.[70] Today, MRSA or hospital 'superbug' remains a major challenge for health professionals. In America alone, for instance, it is estimated MRSA kills 19,000 Americans every year.[71] 'In past viral pandemics, [MRSA] has been the principal cause of secondary bacterial infections, significantly increasing patient mortality rates.'[72] It was, therefore, an added complication during the COVID-19 pandemic, as hospital admissions increased the risk of patients acquiring secondary *Staphylococcus aureus*-related infections. A September 2021 study, for instance, concludes:

> In contrast to patients infected solely with COVID-19, co-infection with COVID-19 and *S. aureus* demonstrates a higher patient mortality rate during hospital admission. [. . .] Our findings emphasize the imperative of COVID-19 vaccination to prevent hospitalization for COVID-19 treatment and the subsequent susceptibility to hospital-acquired *S. aureus* co-infection.[73]

Based on research conducted in a home-made laboratory in his garden at 252 Union Street, Ogston observed, named, and provided the first set of empirical data on *Staphylococcus* – a cocci that still, more than 140 years after its discovery, continues to present extreme challenges to the global medical profession.

In July 1882, Ogston and Bella embarked on what can only be described as an extremely arduous holiday to Norway. The Ogstons sailed from Leith to Kristiansand on Friday the 21st. En route to Stavanger on the 26th, Ogston commented:

> As far as we have yet seen, travelling to and in Norway is very simple, and might easily be done without leaving home. By going to the seabeach, wetting one's clothes, and lying the whole day wrapped in a blanket wrung out of the sea, the conditions & comforts might be very fairly realised. [. . .] The natural consequence is that rheumatism had invaded our frames, and that, unable to move a muscle even to yawn, the daytime is one prolonged torture and the night-time a nightmare.[74]

Added to frequent rain, rough lodgings, and 'unwholesome viands' ('the briny butter, the rancid eel, the mawkish rye bread'[75]) the often egregious state of Norwegian roads was a defining factor in the Ogstons' trip:

> about 2 miles from Hellevik – horror! – The road becomes invisible, though we can here & there track its general direction, but up we go, up slopes of at least 45° with the horizon[,] over boulders, across water ruts, now leaving the horse to its own wisdom and now encouraging or resting the poor animal while we pull our own tired bodies together, then off again, & up & up until

the savagery of despair takes possession of us and in our recklessness we would have charged the mouth of Hell.[76]

Alexander and Bella visited the Buar Glacier, Hardanger Fjord, Sogne Fjord, and the Borgund Stave Church. Recalling the daily routine of 'extrud[ing] unwillingly and lingeringly from our inhospitable couches our sore and stiffened limbs, to encase them in the garments encrusted with the dust and perspiration of the previous day', Ogston claimed this ritual was typically followed, once their horse cart had been set in motion, by a gradual appreciation of 'the balmy warmth of the day anointing our sores & binding up our wounds, till the tired body subsides into a state of tranquil enjoyment of the caress of nature'.[77] Ogston's ability to see adversity as adventure would stand him in good stead as, in future years, he began to replace medical research with a new campaign for improvement: the field of military medicine.

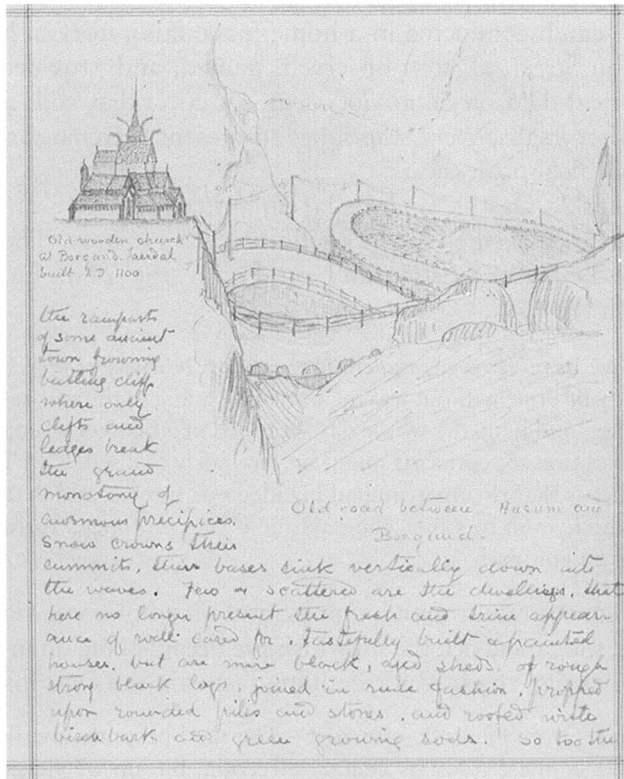

Figure 2.3   Sketch by Ogston of the road between Husum and Borgund, Norway. MS-3850-1-2-00094 in the University of Aberdeen Museums and Special Collections, licensed under CC By 4.0.

3

1883–1899

Military surgery is a subject which has always had a fascination for me.[1]

A S HE ENTERED HIS fortieth year, Ogston continued to make contributions to medical science. In 1884 the *Bristol Medico-Chirurgical Journal* printed his treatment for flat foot, which removed articular cartilage from the surface of the scaphoid (navicular) and astragalus (talus) to re-establish the arch of the foot.[2] Also that year, Ogston published the first description of frontal trephination of the forehead to access the sinuses.[3] And he was Aberdeen University's delegate to the 1884 International Medical Congress in Copenhagen. At that gathering, Ogston conceded, while he was still in favour of his procedure in certain situations, Macewen's operation for *genu valgum* was the better option in most cases.[4] A new front of endeavour was to open up for Ogston, however.

The opening of the Suez Canal in 1869 allowed military and trade vessels a more direct route to the East, removing the need for the lengthy passage around the Cape of Good Hope. By 1882, Britain had assumed virtual control of Egypt and, in order to protect its interests in the region, placed 35,000 troops at Ismailia. Then in response to an uprising at Suakin in Egyptian-controlled Soudan in January 1884, Britain sent out a force under General Sir Gerald Graham. Readers of the 7 March 1885 edition of the *British Medical Journal* were informed that Professor Ogston, having left London on 27 February, was on his way, via Brindisi, to join Graham's troops at Suakin:

> Although it had been intimated by the War Office that no civilian surgeons are required [. . .] Dr. Ogston pointed out that in Aberdeen he had to teach students, many of whom entered the Army Medical Service, and that it was advantageous, for all parties that he should make himself acquainted, practically, with military surgery.[5]

Dr Mackenzie Davidson, who would pioneer a technique for locating foreign bodies in the eye and become a leading figure in the use of X-rays during the First World War, undertook to deliver Ogston's lectures for the remainder of the term, while Ogston, at his own expense, followed Graham's unit. His professed sense of professional obligation in study-ing military medicine in the field is entirely credible; equally, it is hard to disavow the notion the Soudan trip would allow Ogston to exercise his penchant for adventure.

Accompanying the Soudan campaign must have been an enticing prospect for Ogston, for it took him away from an academic battle he had been conducting on the domestic front. The growing medical school at Marischal College struggled to find adequate room to accommo-date its various departments. And, since becoming Regius Professor of Surgery in 1882, Ogston had been incommoded by the lack of teaching space at Marischal College. Calling for the Board of Works to extend the College's north wing in 1883, Ogston complained he could only access a classroom in which to teach surgery for three hours in the morning, as the room had to be vacated for the afternoon session of the pathology professor. The practical class in bandaging and minor surgery, mean-while, took place in a room borrowed from the midwifery professor, and demonstrations for the minor surgery class were held in the patho-logical museum. Furthermore, there was 'nowhere to keep cadavers or for giving the summer class on operative surgery'.[6] Ogston convened a committee to consider large-scale expansion of Marischal College's north and south wings. Eventually, this would see the houses on Broad Street (including the one where Ogston was born – which was adjacent to the College's old entrance) demolished to make way for the ornate grandeur of the current façade, opened in 1906. The University Senate, which approved the extension committee's recommendations in January 1885, could only look at Ogston's trip to Egypt with the most lightly qualified disapproval:

> while fully admitting the unselfish feelings which could have animated Dr Ogston in incurring so much danger and responsibility for the purpose of promoting the interests of science [the Senate] do not feel entitled to sanction under any circumstances the precedent of a professor leaving his duties during the Session without the express approbation of the University Authorities.[7]

Having equipped himself with a Dongolese servant in Cairo, Ogston travelled to Suez and boarded the Suakin-bound hospital ship ss *Ganges*. British forces in the area were suppressing a Mahdist Soudanese army led by Osman Digna. Despite vast British superiority in terms of firepower and organisation, the Mahdists were far from toothless – as Ogston soon saw confirmed by a graphic example. As the ship moored at Suakin, a steam launch brought onboard 'a guardsman transfixed through arm and chest by an Arab spear'.[8] The British encampment – surrounded by a two-foot-high wall – was arranged on the plain adjacent to the island suburb of El-Khaf. Ogston noted the hospital tent

Figure 3.1 View from the door of Ogston's tent in camp at Suakin, 22 March 1882. MS-3850-1-2-00094 in the University of Aberdeen Museums and Special Collections, licensed under CC By 4.0.

THE FIGHT AT HASHEEN, NEAR SUAKIM, MARCH 20: INSIDE THE SQUARE.

Figure 3.2 Ogston providing medical assistance at the Battle of Hasheen. MS-3850-1-3-00002 in the University of Aberdeen Museums and Special Collections, licensed under CC By 4.0.

had been placed in 'the most exposed and dangerous quarter of the camp' and bemoaned the *'sanitary arrangements were only too visible'*.[9] The Principal Medical Officer, Dr Barnett, attached Ogston to the First Bearer Company. Sharing a tent with some of the medical staff, he settled down for the night with a sense of keen anticipation:

> I can even now vividly recall the pure delight, such as life does not often afford, of my first evening out there behind the farthest corner of the low breastwork of the camp, where all was sinking into stillness, fanned by the pure gentle air of the desert on which we looked, watching the purpling sunset sky, ornamented by the horizontally placed crescent of the new moon. [. . .]
>
> Next day promised to be a stirring one, and we were to snatch only a short sleep in our clothes, yet sleep lingered, as I lay opposite the tent door watching the moon drop to the western hills and the stars blaze brighter, till the constant challenge of the sentinels grew fainter and oblivion set in.[10]

Ogston's literary style is typically clear and keenly observational, with an occasional inclination to the ponderous severity of Victorian prose. At times of particular excitement, or in response to exceptional phenomenological beauty, however, his habitually precise empirical style provides a detailed, yet also atmospheric, account of the scenes before him. The frequency of such brilliantly observed moments indicates that Ogston, while viewed by some as undemonstrative and stern, possessed a sensitive inner life.

The following morning, he rose at 4 a.m., buckled on his revolver, and marched with the First Bearer Company as it proceeded in a larger column of some 6,000 soldiers. Having reconfigured into square formation, Graham's troops came under attack: 'The instant the firing broke out our square stopped, and poor L— gave a loud cry and fell backwards off his horse [. . .] shot through the left breast'.[11] Ogston attended 'L' and a major, 'R', who had been 'speared through the right thigh'. (In his sketchbook from the Suakin campaign, Ogston preserved an image of himself kneeling beside Major 'R' which subsequently appeared in an illustrated London newspaper.) Searching for further wounded, Ogston 'witnessed a body of some thousands of Arabs come pouring towards our square from the bush on the north', one which 'by the time they had come within a few hundred yards [. . .] were mostly mown down by the tremendous hail of bullets'.[12]

On the return march to Suakin, Ogston was positioned in the rearguard square. This came under persistent rifle fire and 'began to look somewhat like an ugly business, for the terrified horses, the artillery, and others, hurried forth to the front of the square, leaving its back part empty

[. . .] so that it appeared that as if a breach in the formation were about to follow'.[13] Casualties were heavy and Ogston and the men were 'taxed to the utmost of our capacity' in carrying the wounded.[14] The square was reformed, however, and with field guns having poured shrapnel on the pursuers, the First Bearer Company was able to return to Suakin. Ogston dined on tea and dry bread and 'turned in to sleep [. . .] tolerably tired out by our twenty miles of walking under the tropical sun'.[15]

Ogston also accompanied Graham's troops on an expedition to disperse Mahdist forces from the hutted village of Tamai. During the remainder of his time in the Soudan, Ogston, with the full support of Dr Barnett, studied the methods and organisation of the Army Medical Service. While he praised the dedication and skill of its staff, Ogston identified serious shortcomings. Although medical supplies were adequate, methods of transporting and distributing these were disorganised, while medical orderlies were poorly trained and difficult to discipline. These deficiencies could be rectified, Ogston felt, by granting medical personnel a hierarchy of recognised ranks similar to, and on an equal footing with, the other branches of the armed forces:

> I acquired the conviction that the army medical service could never attain its rightful position until it was put in a place of honour at least equal to that of any other branch, until it was recognised that the education and training of its officers entitled them to be ranked at least equally with the other officers in the fighting forces, [. . .] and until in fact our nation had come to understand that it is not less important to save the lives of its battle worn men and officers than to destroy those of the enemy.[16]

He concluded of his journey to Soudan:

> one of the results of the experiences I gained at Suakin was a resolution which I formed that, if it should ever lie in my power, I should strive to advocate such improvements of the services [. . .] without regard to the odium which is the sure portion of every one who ventures to suggest reforms in the War Office.[17]

Ogston was awarded the Egypt Medal and the Khedive's Star for his role in the Soudan campaign. The 31 October edition of the *BMJ* reported 'the return of Professor Ogston from Soudan' had precipitated 'some talk [. . .] about the formation of' a volunteer medical staff corps at Aberdeen University.[18] In 1883 James Cantlie (a Banffshire native and MB CM [1873] graduate of Aberdeen University) established a volunteer medical staff corps from pupils at Charing Cross Hospital, London. And in November 1885, with a view to a similar initiative being started in Aberdeen, Cantlie was invited to address a meeting of the city's medical students. He was warmly received, and the students resolved to form such a body. The Aberdeen Medical Staff Corps was duly established in 1889 – the same year it was reviewed, during a training exercise in Aldershot, by Kaiser Wilhelm II.[19]

Frictions with the university continued after Ogston's return from the Soudan. Upon recommendation of the General Medical Council, the university decided to move the surgery examination from the third year of study to the final, fourth year. In the transition period, Ogston stood to lose a year's income. With intransigent resistance to such treatment, he 'found it impossible to maintain relations with members of Senatus and the Faculty of Medicine, and declined to serve any longer as convener of the Committee of Extension, promoter to the degree of CM, or on the Finance Committee'. The Senate thereafter withdrew its support for relocating the surgical examination.[20]

Despite such frictions, Ogston was evidently held in high regard, within and outwith his profession. In 1896 he was made a Deputy-Lieutenant of Aberdeenshire and, in August 1887, was appointed Vice-President of the Surgical Section of the British Medical Association's annual conference in Dublin. The following month, on 25 September, his father, Francis, died in Aberdeen. In 1888, in a development which would further consolidate the position he now occupied as head of the family, Ogston purchased the estate of Glendavan in Deeside. Located thirty miles west of Aberdeen, Glendavan is situated to the north of Loch Davan, a small body of water immediately north of the much larger Loch Kinord. Together, these form part of the Muir of Dinnet nature reserve. On a hill overlooking Loch Davan, Ogston built Glendavan House. The original design comprised six bedrooms, a night nursery, day nursery, and schoolroom, with additional accommodation for servants and the family nurse, Mrs. Maxwell. To reach Glendavan, the Ogston family would have travelled on the Aberdeen–Ballater railway line as far as the village of Dinnet, before taking a horse and buggy the remainder of the journey to Glendavan. The house provided a gathering point for immediate and extended family. Here, the Ogstons and their guests fished, shot game, played croquet, and explored the surrounding countryside.

Ogston's leisure time was characterised by the same exactitude that shaped his professional life. Walter notes: 'Father's shoots at Glendavan were, like all his other doings, arranged and planned each day so that there was no waste of time [. . .]. Each man was allotted his place in the line of guns, which he kept to for the whole of the morning or afternoon.'[21] Ogston decided the daily activities to be undertaken by those staying at Glendavan; however, despite this somewhat regimented approach to holidaying, the house was a site of many happy memories for the Ogston family. Walter recalled fondly:

> The end of each day at Glendavan was generally marked by what I may almost call a ceremony. This was the closing of the gate at the entrance to the avenue. Most of us would troop down with [Ogston] in the dark to do this, and if the sky was clear, there would be some star gazing, picking out the chief stars and constellations. Occasionally, on very fine moonlit nights, we would cross the road into the boathouse wood, and look across the loch from under the trees, and admire the spread of the water with its beds of reeds along the sides.[22]

Ogston built a cottage, Langcroft, near Glendavan for his daughter Mary and her family. Mary's daughter Molly recalled with delight: 'All August we lived there in complete satisfaction; we lived almost entirely in the open air; and no words can express how beautiful that countryside is, or how we loved it.'[23]

In 1892 Ogston generated widespread alarm by announcing his – apparently totally unexpected – decision to retire from Aberdeen Royal Infirmary. The 23 April edition of the *BMJ* reported Ogston, then approaching his forty-eighth birthday, 'sent a letter to the chairman of the Board of Directors of the Aberdeen Royal Infirmary, resigning his appointment as senior surgeon'. A meeting of infirmary staff carried the following unanimous resolution:

> That this meeting regards Dr. Ogston's withdrawal from the staff of the Aberdeen Royal Infirmary as a serious and almost irreparable loss to the institution, to the University, and to the general community, and that it pledges itself to use every effort to induce Dr. Ogston to continue in the office which he has so long conspicuously adorned.

A second resolution, proposed by Dr Profeit of Balmoral, was passed to encourage the ARI directors to do all in their power to retain Ogston as Senior Surgeon. At Marischal College, meanwhile, a large body of students gathered to bemoan the sudden and dispiriting news of Ogston's intention to retire. Here, likewise, a resolution was passed 'begging him to reconsider his decision', while a 'deputation of students was appointed to lay this resolution before Dr. Ogston'.[24]

At a meeting of the ARI board of directors, the chairman read Ogston's letter, where he claimed he wished to retire because: 'My University duties and private work, especially at certain seasons, occupy a larger portion of my time than formerly, while the work of my department of the infirmary has increased to a degree that renders it beyond my power to do it justice.'[25] (Ogston had previously been in dispute with ARI colleagues concerning curtailment of his access to its sole operating theatre.[26]) The directors professed themselves eminently desirous of retaining Ogston, and offered him a six-month leave of absence. On 21 May, the *BMJ* ran a note stating Ogston had written to ARI and was willing to withdraw his resignation 'if certain points be granted him'.[27]

Having a country home on the Aberdeen–Ballater railway line was, Ogston thought, a factor which likely contributed to the next phase of his career. In 1892 Ogston was made Surgeon-in-Ordinary in Scotland to Queen Victoria. 'It was all so unexpected, and I was so unprepared for it,' Ogston wrote. 'I was but an obscure provincial surgeon, whose dreams never rose above being a moderate success.'[28] Though becomingly modest, this is simply not true. By the age of forty, Ogston was Senior Surgeon, Regius Professor, and a leading researcher in the understanding of pathogenic bacteria. Moreover, he was, as he later learned, nominated for the royal post by Sir William Jenner, the Queen's Physician in England, and chosen over Hector Cameron, Joseph Lister's former house surgeon and assistant at Glasgow University.[29] Ogston received the offer of the Surgeon-in-Ordinary position in a letter sent to him by Dr (later Sir) James Reid (1849–1923). Initially Queen Victoria's Resident Medical Attendant, Reid would later become her Physician-in-Ordinary, in a role that evolved to extend far beyond medicine, when Reid became one of the queen's closest confidantes.

From Ellon, Aberdeenshire, Reid took up his post in 1882 and aptly fitted Victoria's desire for a medical attendant from Aberdeenshire who spoke German (enabling Reid to communicate with the queen's relations). Ogston, likewise, knew the German language and came from the north-east of Scotland. These facts, and Ogston's residence at Glendavan, may have suggested him to Reid as a suitable candidate. Practical considerations aside, Victoria was deeply attached to Deeside. She and Prince Albert purchased the Balmoral estate in 1852, where Balmoral Castle was completed three years later. Following Albert's death in 1861, Jenner suggested Victoria's grief might be alleviated by John Brown (1826–83) – who was born on the Balmoral estate and had served as Albert's ghillie – becoming her groom. Subsequently, Brown's role grew into that of a much trusted and valued personal attendant of the queen. It is possible, therefore, that, in addition to his international medical reputation, Ogston was – as an Aberdeenshire man and one appreciative of the charms of Deeside – well suited to become surgeon to a monarch with strong emotional ties to the north-east of Scotland.

On 20 September, Ogston took the train to Ballater, where he travelled by coach to make a courtesy call at Balmoral Castle. Although the queen was in residence, Ogston had no inclination he would be seen by the monarch that day and so was attired 'in my ordinary working dress, wearing a coloured shirt and having on a pair of old shoes'.[30] Ogston met the queen's commissioner at Balmoral, Dr Profeit, and her private secretary, Major Bigge. Upon learning from Bigge that Ogston was at the castle, Victoria intimated she desired to meet him after lunch. This news threw Ogston into a fit of uncharacteristic panic. He recalled:

> I have lectured before hundreds of the most famous men in the world; I have done, for the first time, great and dangerous operations; I have seen men shot down at my side; have enjoyed my luncheon and pipe when bullets were rattling all around [. . .]. But to know that in a few minutes I was to stand before the Empress of Britain and India [. . .] made my heart sink within me.[31]

Ogston lunched with members of the royal staff, drinking Hungarian Ruster wine, and – incongruous though it may seem – 'detailing the literature on the subject' of sea snakes for the benefit of his companions. Ogston remembered, 'All through it I had the strange impression of being a dual personality: one of an unruffled individual quietly talking away, and the other of a condemned felon counting the minutes preceding his execution.'[32] A somewhat comic situation developed. Of the moment he was told the queen was ready to see him, Ogston remembered: 'I then knew what it must feel to be a jelly-fish.'[33] When admitted to the queen's presence, Ogston took meagre consolation from his proximity to a cabinet which he could 'lean back against [. . .] and so prevent myself from fidgeting'.[34] In conversation, however, Ogston was unable to do anything more than bow and distractedly mumble responses to the queen's polite questions. Perspiring, Ogston noticed to his horror there was a hole in the top of his shoe, and felt the impulse, as he recalled, 'to laugh aloud from the vexation I felt'.[35] He was then led from the room by the queen's secretary.

Queen Victoria, if she perceived any embarrassment, made no note of it in her journal, which noted simply: 'After luncheon saw Professor Ogston of Aberdeen, just appointed my Surgeon in Scotland.'[36] Having survived this initial encounter, Ogston continued to serve the queen and her family for several years, retaining his position as Surgeon-in-Ordinary under Edward VII and George V. He formed a high opinion of Victoria, writing: 'Her whole conversation and bearing were absolutely natural, with no touch of pride or haughtier, perfectly human, womanly, and sweet. I felt that she was a most intelligent person, with a marvellous memory, and a genius for detail.'[37] Ogston did not wish for accounts of his professional dealings with the royal family to become public; therefore, details concerning his role as royal surgeon are scant. Walter hints at the scale of his father's role, remarking that Ogston 'had a number of interviews with Her Majesty' and that 'these interviews were concerned with intimate matters relating to various illnesses of members of the Royal Family or their servants'.[38]

In 1893, however, Ogston attended a levee at St James's Palace, London, presided over by the Prince of Wales. The following year, he received the Duke of York at Aberdeen Royal Infirmary, which the latter visited in his capacity as President of the Highland and Agricultural Society. Moreover, Ogston became close enough to Dr Profeit for Walter to describe the queen's commissioner at Balmoral as 'an old friend' of his father, who 'invited [Ogston] to fish for salmon or to shoot hinds [. . .] when the Royal Family was not in residence at the Castle'. Occasionally, Walter was allowed to accompany his father on these 'memorable' occasions.[39] Ogston's role as Surgeon-in-Ordinary, then, was more than titular. It brought him into contact with the royal household and – as would later become apparent – was in terms of his relationship with Victoria, highly advantageous when Ogston continued to pursue his investigations into British military medicine.

1894 also saw Ogston make a voyage to South Africa. On 23 August, he sketched a view of the Canary Islands and, having passed St Helena, caught his first glimpse of South Africa on 8 September. This journey was possibly to visit his estranged son Francis (1869–1901). As Janet Teissier du Cros describes events, Francis had been sent to Clifton College, Bristol, for his schooling and, having no pocket money of his own, had stolen from a schoolfellow. Upon his expulsion, Francis returned home and 'was turned away by his father and stepmother'.[40] Thereafter, he worked for Messrs George Thompson and Co. and lived in India and South Africa. Ogston, however, does not appear to have left any diary record of the 1894 trip other than a sketchbook, which indicates his journey took him to the Karoo Desert near Matjiesfontein, then Kimberley, where he watched the case of Thompson v. Oliver in the Kimberley Court House on 13 September. The same day, Ogston made a sketch of 'Mr Justice Solomon's Bungalow'.[41] This, as he later elaborated in his Boer War journals, was the home of Sir William Solomon (later Chief Justice of South Africa from 1927 to 1929), a cousin of Bella.[42] Ogston also took in the Paarl Rock near Cape Town on 19 September, and on the 28th, he was sketching flying fish found off the coast of Sierra Leone as he sailed back to Britain.

In 1896 a prestigious son-in-law entered the family when Herbert Grierson married Ogston's first daughter, Mary. Born in Lerwick, Grierson (1866–1960) was educated at the universities of Aberdeen and Oxford and appointed as Aberdeen University's first Professor of English Literature in 1895, before taking up the role of Professor of English at Edinburgh University in 1915. In the course of his influential career, Grierson would do much to promote interest in the metaphysical poets, particularly John Donne. He was one of the first university professors to offer a modern English literature course in Britain, providing an over-view from ancient literature to near contemporary works, which became the standard model for the subject until the 1970s. Grierson first encoun-tered Ogston's daughter Flora at a dinner in Aberdeen and thought her 'strikingly handsome' and later met Flora and Mary at dances in the city. Grierson was attracted to both sisters: 'if Flora were the more immedi-ately attractive, Mary had the deeper insight, [and] would have under-stood my interest in poetry and my work at the University'. However, 'it was not easy to get to know either of them at all intimately' for '[n]o one visited lightly at 252 Union Street'.[43]

Fortunately, Grierson's role at the university put him in contact with Professor Henry Cowan and his wife (Ogston's sister) Jane. Through this connection, Grierson got to know Mary, and the couple became engaged in June 1896. That year, Ogston had served as President of the Surgical Section of the BMA's gathering at Carlisle. Having been invited out to Glendavan in August, Grierson found Ogston 'revising, I think for publication, the address he had delivered'. Ogston 'was a little taken aback when' Grierson 'began to criticise it closely not from the point of surgery but of Rhetoric and English'. Rather than accept any of Grierson's specific recommendations, Ogston merely marked the deficient areas with 'a blue pencil' and by 'next day had rewritten the sentences'. Grierson summarised Ogston: 'He was, I suppose, a severe man, and could check small things in a way that made you sit up. But in bigger matters, even if he thought you had been foolish, he thought only of how to help you.'[44]

Some thirteen years following the Soudan, Ogston continued to take an interest in the infrastructure of British military medicine. He 'repeat-edly visited Germany' to study arrangements there and '[a]rmed with recommendations from Queen Victoria' made inspections of 'the leading English military hospitals'.[45] Victoria, indeed, had long taken an interest in military medical arrangements. During the Crimean War, she visited the military hospitals at Fort Pitt and Brompton and, in 1856, laid the foundation stone of the Army Medical School's hospital at Netley. After one such hospital inspection, she wrote to the Secretary of State for War, Lord Panmure, to recommend, regarding:

> hospitals for our sick and wounded soldiers [. . .] This is absolutely the time to build them, for *now* is the moment to have them built, for no doubt there would be no difficulty in obtaining the money requisite for this purpose from the strong feeling now existing in the public mind for improvements of all kinds connected with the Army.[46]

Victoria's personal investment in improving military medicine may have been of no small benefit to Ogston, enabling him to make his enquiries with her official sanction and approval.

In his endeavours, Ogston was furthering the work of a distinguished Aberdonian medical predecessor, Sir James McGrigor (1771–1858). McGrigor was Inspector-General of Hospitals during the last two and a half years of the Duke of Wellington's Peninsula Campaign (1808–14) against Napoleon, and he later served as Director-General of the Army Medical Corps from 1815–51. McGrigor's segregation of infected soldiers, reduction of hospital overcrowding, and efficient administration earned him the sobriquet 'The Father of British Army Medical Services'. McGrigor's high standards were not, however, maintained during the Crimean War (1853–56). The Army Medical Department only had 163 surgeons and two ambulance wagons (which were left at a Bulgarian port). Its stretcher bearers, meanwhile, were elderly and untrained men from the hastily extemporised Hospital Conveyance Corps, few of whom actually reached the front.[47] Moreover, overcrowded hospitals and poor sanitation led to high numbers of soldiers dying from disease.[48]

A Royal Commission was held in 1857 to investigate the evident failings of the army's medical arrangements during the Crimean War. As a result, a new Army Medical School was established at Fort Pitt in 1860. It was later relocated to Netley in 1863 when the hospital there opened. Additionally, the Army Hospital Corps (renamed the Medical Staff Corps in 1884) was established to provide orderlies and stretcher bearers; however, the Hospital Corps did not have its own officers and was hampered by a necessary lack of leadership and discipline.[49] In 1873 the regimental hospital system (where each regiment had its own medical officer and hospital) was abolished; instead, resources and manpower were reorganised in a modern system of stationary and garrison hospitals, with medical officers now operating under the aegis of the Army Medical Department.[50] Medical officers, however, did not hold ranks equivalent to their military counterparts. So unfavourable was the prospect of a career in the Army Medical Corps, the Army Medical School at Netley had no new recruits in 1878 and 1879.[51]

In 1898 Ogston formed part of a deputation, alongside Sir Thomas Grainger Stewart (President-Elect of the British Medical Association) and Surgeon-General James Mouat VC, who sought to resolve these long-standing deficiencies in the infrastructure of military medicine. This high-powered delegation was received at Lansdowne House on 20 January by Lord Lansdowne, Secretary of State for War. As a result, on 23 June a Royal Warrant was issued announcing the formation of the Royal Army Medical Corps. This new body was established through merging the Medical Staff Corps and the Army Medical Department. Meanwhile, RAMC officers, it was announced, would bear the same military rank – up to the level of colonel – as officers in the army.

In July 1898, Ogston became Medical Referee of the Sheriffdom of Aberdeen, Kincardine, and Banff for the Workmen's Compensation Act of 1897, an Act entitling employees to compensation for a workplace accident that was not their fault. Then, on 1 September – thirty years after joining as Ophthalmic Surgeon to Aberdeen Royal Infirmary in 1868 – Ogston resigned as Senior Surgeon. November, meanwhile, saw him receive an invitation from the Council of the British Medical Association to deliver the Address in Surgery at the Association's annual conference, to be held in Portsmouth the following August. As the content of Ogston's speech would demonstrate, the formation of the RAMC left him far from satisfied with contemporary standards of British military medicine.

At the close of 1898, Ogston extended his military medical investigations to Russia, where he attended the centenary celebrations of the St Petersburg Imperial Academy of Medicine. Ogston left Aberdeen on Christmas Day at 3.30 p.m. and took the sleeper train to London. Crossing the Channel, he travelled to Eydtkuhnen (modern day Chernyshevskoye in Kaliningrad Oblast). Having to spend the night there, Ogston took a room at the inn and got into bed '[w]ith no little trepidation [. . .] for I smelled, or fancied I did, the old familiar scent of bugs'. To mitigate against possible attack, Ogston 'put in practice an old plan that used to do some good. Lighting all the candles & supplementing them with my own stock, I arranged them around the bed and lay like a corpse laid out in state'.[52] He slept from 3 a.m. to 5.30 a.m., woke unscathed, and carried on to Russia. En route Ogston was served well by his characteristically thorough preparations for the journey:

> Even to have learned the Russian alphabet and numbers, as I did, is a vast advantage to a traveller. He can read the names of the stations, the smoking

or other carriages, & the simpler notices, which are really the most a traveller by rail requires. Without the alphabet they are utterly indecipherable.
He can't even find the eating room.[53]

A fellow passenger was able to assist Ogston's transfer to a faster train.
Ogston 'had a long talk' with this man and told him a story about
Donald Stewart, a keeper at Balmoral, asking Tsar Nicolas II: 'Faur's
your gun, min?'[54]

Having settled into St Petersburg's Victoria Hotel, Ogston attended
receptions at the Imperial Academy, the Salle de la Noblesse, and
inspected the medical factory of the Russian War Department. Then,
on 4 January 1899, he was presented to Tsar Nicholas II. The meeting
took place at 'a large riding school in connection with the school for the
training of young military cadets'.[55] Despite his ostensible indifference to
individuals of high rank, the profuse detail with which Ogston recorded
this day belies a high level of earnest excitement. He was not unmoved
by the procession of officers, cadets, and ecclesiastical dignitaries: 'Such
a display of uniforms as passed up before us I never saw [. . .] uniforms of blue, sashes of pink & blue, orders, medals, decorations: it
was extremely pretty but indescribable as a kaleidoscope.'[56] The Tsar
was a distant relative of Queen Victoria and married a granddaughter
of hers, Alexandra of Hesse. Ogston's mutual connection to the British
royal family may have been of some comfort to Ogston as he waited to
exchange a few words with the Russian Emperor. Eventually, the great
moment arrived:

> We were placed in line[,] hats and gloves on, and the Tsar came along and
> shook hands and spoke with each of us in turn, commencing on the right,
> and with each in his own language – excepting, I suppose, the Swedes, etc.
> English he spoke perfectly, better than some of our royal princes do. [. . .]
> His manner was absolutely natural & frank, & had no trace of haughtiness
> whatever in it [. . .].
>
> To me the Czar spoke of having been at Aberdeen, and Balmoral, and I
> told him we hoped he would come back.[57]

Ogston returned to Scotland on 6 January. Later in the month, he received
note from Lord Salisbury informing him the queen had approved of his
accepting the St Petersburg Imperial Academy of Medicine centenary dec-
oration. In March, meanwhile, Ogston was informed he had been made
an Honorary Member of the St Petersburg Medical-Military Academy.

In May 1899, in preparation for his speech, Ogston visited the
Portsmouth docks and warships and the Netley hospital. Then, on
3 August, he delivered his Address on Surgery at the BMA's annual

gathering. Though courteous, Ogston's speech unequivocally and comprehensively laid bare the alarming defects in the army and navy medical services. He began by pointing to the huge advances in elective surgery made possible by Lister's introduction of antiseptic surgery. The 'condition of our Army and Navy Medical Services,' Ogston contended, had, since Listerian innovations became widespread, been 'gradually falling into a condition that cannot fail to evoke much anxiety regarding their present and future'. Ogston noted although Britain had been victorious in West Africa, the Nile region, and the Himalayas, our 'enemies there are poorly armed and our casualties few' and 'it dare not be inferred from this that we are in a condition to deal with a war against a Great Power, with arms and organisation at least equal to our own'.[58]

Ogston recalled the BMA's delegation to Lord Lansdowne the previous year and quoted the words of 'an eminent President of this Association' (presumably Sir Thomas Grainger Stewart, elected BMA president in 1898) who had cautioned: 'If there were to come a time of war it is to be dreaded that the horrors which occurred during the Crimea may be repeated.' These sentiments having been echoed by others present at the January meeting, the Secretary of State had conceded 'that, owing to the condition of the Army Medical Department, "only comparatively inferior men present themselves . . . . . . the best men turning away from the Army Medical Staff . . . . . . a very grave condition of things"'.[59]

Specifically, Ogston pointed out, since no major British campaign had been staged since the introduction of antiseptic surgery, there was urgent 'need for a class of surgeons more highly trained than hitherto to carry the best methods of antiseptics into the domain of gunshot wounds'. However, '[u]nder the existing system no junior officer in the Royal Army Medical Corps has any opportunity of practising modern surgery'. A state of affairs in which 'junior medical officers have neither the chance nor the encouragement to perfect themselves in their science' was, Ogston opined, one of 'intellectual sterility' from which 'our best graduates recoil'.[60]

Although scientific developments were incorporated into other branches of the military, such as engineering, artillery, and telegraphy, and their men rigorously trained to a high state of efficiency, medical matters were 'left in a condition which may have answered well enough fifty years ago, when rough-and-ready surgeons could do the rough-and-ready work then required, but which is out of harmony with the revolutionised conditions of modern surgery'.[61] Ogston felt medical forces should be trained constantly, their staff kept up to date with literature on medical developments, and be in possession of the most modern equipment and instruments. As he summarised, in the armed forces, everything 'that can destroy life is in the highest perfection', while the means which 'can save it is of the rudest description and behind the age'.[62] By way of contrast, Ogston pointed to Germany, where army surgeons were sent 'to do duty for a year at a time in the medical, surgical, and gynaecological departments of the hospitals connected with the universities' and to attend 'special courses of instruction in anatomy, surgery, and operative surgery'.[63] Russia, likewise, presented an example worthy of emulation, possessing '6 large and 300 small hospitals connected with the army', where 'in some of these the equipment is far in advance of anything the services possess in this country'.[64] France, furthermore, provided postgraduate courses expressly for army medical doctors, who were offered extra pay to attend them.

In terms of recommendations for reform, Ogston urged that the Royal Army Medical Corps be rearranged to meet its ostensive ability to function as three complete army corps (a system which 'if it exists at present, is a mere framework on paper'). Ogston urged:

> The number of those who are to serve when war breaks out should be completed; the equipment and materials should be ready and at hand; and men and officers frequently trained in their use. Field hospitals should have their equipment complete, and their staff attached to them. And the whole service should be organised as a body of independent units, capable of acting alone or being combined or brigaded as required. Everything belonging to each unit, even to its transport, should be complete and under the command of its medical officers.[65]

Furthermore, Ogston strongly urged medical officers be granted study leave in order to keep themselves familiar with contemporary advances in medical science. He recommended expanding the staff of military and naval hospitals and opening these institutions to civilian invalids, to provide opportunity for continual practice by the medical branches of the armed forces. Finally, addressing the notion that medical aid in wartime could come from the civilian sector, Ogston opined that '[n]o more erroneous idea could possibly exist'.[66] Voluntary civilian aid, 'even such as that proposed by the Central British Red Cross Committee' could not be adequate, unless it were expanded and developed to such a level as to render the RAMC superfluous. The pernicious hope that civilian aid could replace, or efficiently supplement, the military's medical infrastructure was one that had 'already done infinite harm', and an equivalent civilian intervention in the work of the Royal Engineers, for instance, would never be countenanced.[67]

Astutely, Ogston foresaw the British Army – its medical provisions already exposed as inadequate in minor colonial campaigns – would 'have its eyes rudely opened' in the event of a larger, more demanding conflict that would expose 'how unfitted the services are for their great and real work'.[68] Surgeon-General Robert Harvey commended Ogston on his speech, 'though some parts of the address would no doubt stir the profession and cause a certain amount of criticism'.[69] The anticipated opprobrium followed in abundance. On 23 September, the *BMJ* printed a notice which acknowledged reforms were required in the medical arrangements of the armed forces, but countered that Ogston had, 'in other directions over-stated his case, thereby not only endangering the reforms desired, but arousing needless irritation in both services'. The heated nature of the dissatisfaction generated is indicated by the author's claim Ogston 'indulged in comparisons which, generally needless, often injudicious, and always dangerous, even if possessing an element of truth, are as a matter of policy better avoided'. Moreover, it was claimed, 'the address has been made a peg, both on the platform and in the press, on which to hang gross and outrageous statements by those who seem to take an unaccountable delight in vilifying medical officers'.[70]

The events of the Boer War (1899–1902), however, reassured Ogston that the urgency and severity of his criticism had been merited: '*Within a very few months after my address, and while it was still fresh in men's minds, the South African War broke out*, and all that I had asserted in the Portsmouth Address was amply confirmed.'[71]

4

1899–1900

My Dear Professor Ogston, – The Queen is much pleased and interested in hearing of your intention to visit South Africa, for the purpose of studying military surgery in the operations of our troops.[1]

THE BOER WAR (1899–1902) was the result of long-standing tensions in South Africa between the Boers (settlers of Dutch origin) and the British Empire. The 1881 Convention of Pretoria granted the Boers control of domestic affairs, while Britain maintained suzerainty and managed foreign affairs. In 1886, however, relations were destabilised by the discovery of gold in the Witwatersrand, leading to an influx of – mainly British – 'Uitlanders'. In 1897 Alfred Milner – High Commissioner for South Africa and Governor of Cape Colony – pressed the Boers to alter their constitution to give greater rights to British Uitlanders. Unable to gain the desired concessions, Britain sent troops to strengthen its position in South Africa. Conflict broke out when the British ignored a Boer ultimatum that a state of war would exist unless the additional British forces were withdrawn.

As Nicholas Murray outlines, the Boer War marked a significant development in the strategies of modern warfare. It saw the first mass-scale use of small-bore magazine rifles, trench fortifications, and barbed wire. The firing of heavy artillery, meanwhile, rendered close-order marching obsolete and meant troops had to operate in open-order formation. Having defeated the main Boer forces in spring 1900, the British fought the remainder of the conflict against a depleted foe now adopting guerrilla warfare tactics. Pursuing their enemy over a vast terrain, it took the British nearly two years to force the Boer surrender in May 1902. British victory was achieved by sending in tens of thousands of additional troops and by the rounding up of Boer women and children into concentration camps – a strategy adopted to comprehensively rout out Boer forces and support networks, on an area-by-area basis.[2]

One particular aspect of contemporary warfare which piqued Ogston was the issue of Dum-dum or 'exploding' bullets – a matter which generated international controversy prior to and during the Boer War. In the St Petersburg Declaration of 1868, the concerned nations issued the following statement intended to prohibit the use of bullets which would explode on impact with the human body:

> The Contracting Parties engage mutually to renounce, in the case of war among themselves, the employment by their military or naval troops of any projectile of a weight below 400 grammes, which is either explosive or charged with fulminating or inflammable substances.[3]

Subsequently, however, Britain developed the Dum-dum bullet for use in colonial wars. During the Chitral campaign, rounds fired by the Lee-Metford rifle and its successor, the Lee-Enfield (adopted by the British Army in 1895), were found to be insufficient at immobilising enemy troops.[4] It was discovered, however, that by removing a fraction of the bullet at its apex, a round would expand on impact, generating a more deadly wound. The name 'Dum-dum' arises from the location of the arsenal in Calcutta where these bullets were manufactured. Then, in 1897, the British adopted the hollow-pointed Mark III bullet, which was soon superseded by the more deadly Mark IV, a higher velocity round with 'a three-eighths inch cylindrical hole punched in its tip'.[5]

The 28 May edition of the *BMJ* printed a letter of Ogston's where he related hearing, at the April 1898 Congress of German Surgeons, Professor Von Bruns's description of Dum-dum bullets as 'inhumane projectiles'.[6] Von Bruns's findings led to the Congress condemning Dum-dum and Mark IV bullets and calling for an emendation to the St Petersburg Declaration that would ensure only bullets which were wholly mantled, or mantled at the apex, would be used.[7] While Von Bruns had not yet published his research, Ogston had discussed it privately with him. Thus informed, Ogston raised the objection Von Bruns had, in fact, conducted his research using Mauser big game rounds (which were 'stripped of their mantle for about 1 centimetre from the tip') in contrast to the Dum-dum (which 'has its mantle removed for only 2 or 3 millimetres from its tip'). Consequently, Ogston remarked, the 'explosive or expanding action must be much less marked' in the Dum-dum than in the rounds Von Bruns had actually studied.[8] Ogston developed these observations in a *BMJ* article published in September 1898, where he provided a more detailed treatment of the issue. Dum-dums, he clarified, belonged to a class of small arms round he called a 'Disintegrating' bullet: 'Soft leaden bullets or mantled bullets with leaden tips, [which] when the velocity and energy are great tend to fly in pieces or disintegrate on impact'. These he distinguished from the 'Explosive' bullets – 'provided with substances that detonate on impact' – which were prohibited by the St Petersburg Declaration.[9]

Ogston devoted considerable energy to the issue – partially, at least – because he believed criticism of Dum-dum bullets represented a genuine threat to British colonial campaigns. Not only did he dispute the validity of Von Bruns's conclusions, Ogston also asked if it was appropriate to adopt the most destructive weaponry available if the enemy in question was bent on conducting war without any humanitarian regard for its foe. His phrasing of this argument is shaped by what are, inescapably, white supremacist sentiments. In a March 1899 edition of the *BMJ*, Ogston queried whether:

> what is unreasonable in contending with a civilised foe who gives quarter and cares for the lives of wounded, sick, and disabled men, is also unreasonable when dealing with those who, if successful, wage a war of annihilation, and dispatch armed and unarmed, wounded, sick, men, women, and children alike, whether on the battlefield or in cold blood, and they be overpowered in war or captured by stratagem or treachery?
>
> These are some of the questions we ought to consider and be prepared to answer, since upon our reply to them may depend, for instance, whether it will be possible for our soldiers to avert an impending disaster that will entail their annihilation, or even, it may be, lead to English men and women falling sacrifices to some African juju, in some African city of blood, by the use of small-arm projectiles more efficient than they would employ against a civilised nation, kindred, perhaps, to ourselves and blood.[10]

There is some inconsistency between these remarks and Ogston's comments delivered later that year at the BMA's conference at Portsmouth in August, where he noted that, in West Africa, the Nile region, and the Himalayas, Britain's 'enemies there are poorly armed and our casualties few' – and not to be compared to the military might of a 'Great Power, with arms and organisation at least equal to our own'.[11] Ogston's remarks can be understood in the light of coming from a very different time – one where patriotism was synonymous, for some, with whole-hearted belief in maintaining the British Empire. Nonetheless, the disturbing implication here is not only is there such a thing as more and less civilised human beings, but that it is advisable to reserve the most destructive weaponry for use against less militarily developed nations.

Ogston was also concerned, however, that international condemnation of Britain's use of Dum-dum bullets extended not only from moral outrage formed on erroneous data, but from a desire to place Britain at a military disadvantage. At the May 1899 First Hague Conference, twenty-three of the nations represented voted in favour of Declaration (IV,3) concerning Expanding Bullets:

> The Contracting Parties agree to abstain from the use of bullets which expand or flatten easily in the human body, such as bullets with a hard envelope which does not entirely cover the core or is pierced with incisions.[12]

Only the United States and Britain voted against the motion, with Portugal abstaining. Having heard reports of this declaration, Ogston concluded – given the delegates' apparently deliberate willingness to ignore the inapplicability of Von Bruns's findings to Dum-dums – that the Hague Conference Declaration was, in fact, an attack on the efficacy of the British military:

> an endeavour to coerce her into an undertaking to make no attempt to render her military rifle equal to theirs. Whether or not this was their real motive, things have been so injudiciously handled as to give rise to the suspicion.

Ogston pointed out that the weight and velocity of a bullet, as well as its propensity for fragmentation upon impact, were its key destructive attributes. He poured scorn on the implication that a fully mantled bullet 'lets the life out of a man by the smallest opening possible and in the gentlest way – a utopian sort of bullet'.[13] Ogston argued that all modern bullets could cause grievous suffering and described all such projectiles 'as used by man against man' as 'a disgrace to our species'.[14]

By conducting experiments with cadavers, Ogston demonstrated the difficulty of obtaining the kind of reliable data which would be required before any official banning of Dum-dum bullets could be scientifically validated. Firing three types of British bullet (fully mantled, Dum-dum, and IV Woolwich) and taking wax casts of the internal cavities produced by their impacts, Ogston concluded 'that in flesh wounds the three bullets do not produce such marked differences either in openings or tracks as has been asserted'. Furthermore, by firing against a human arm from a distance of six feet, Ogston showed that the Mauser bullets employed by Von Bruns produced a much larger exit wound (4.5 by 2.25 inches) versus that made by the Dum-dum (2.75 by 1.5 inches).[15] Ogston provided photographs of these exit wounds and of the bone fragments collected from each, thereby demonstrating the Mauser big game rounds resulted in significantly greater fragmentation of bone.

Ogston's findings corroborated his assertion that experiments with Mauser bullets could not be used to assess the destructive capacity of Dum-dums. Therefore, the matter of Dum-dum bullets was 'by no means so plain and simple as has been represented', but rather had been discussed in a manner 'illogical and unscientific, to say the least of it'.[16] Moreover, the Declaration itself was worded in unhelpfully vague and inaccurate language, since none of the signatory nations used bullets which possessed a mantle 'completely cover[ing] the leaden core'. This, Ogston argued bluntly, was impossible, since: 'There must always, wherever it be placed, be an opening in the mantle through which the leaden core can be introduced.' Accordingly, the Declaration was 'an absurdity, condemning the fully mantled bullets used by the Powers who drew up the report quite as much as it does the Mark IV or the Dum-dum bullets'.[17]

Regardless of the scientific deficiencies Ogston identified in the Hague Declaration, as Britain and the Boer Republic were not signatories, they were not obliged to adhere to the convention's prohibition of expanding or non-fully mantled bullets. And some 66 million Mark IV bullets had been produced by March 1899. The British government, however, was wary of damaging public opinion, and '[a]lthough the Dum-dum cartridge was still used in other parts of Africa and Asia, it was deemed unfit to use against white adversaries' during the Boer War.[18] Ogston went to the Boer War, then, not only as a high-profile critic of the British Army Medical Service, but as a well-informed participant in the public debate around the use of expanding ammunition in the build-up to the conflict.

Following his visit to Balmoral Castle on 8 November, Queen Victoria – having either heard directly from Ogston or through an intermediary – approved of his plans to accompany the British campaign and 'promised introductions to the military authorities there'. On 24 November, Arthur Bigge sent Ogston the following note:

> My Dear Professor Ogston, – The Queen is much pleased and interested in hearing of your intention to visit South Africa, for the purpose of studying military surgery in the operations of our troops; and Her Majesty highly appreciates your self-sacrifice in thus at your own expense going to the seat of war, with the intention of giving your valuable services to the wounded.
>
> By Her Majesty's command, I am writing to the military authorities to render you all possible assistance.

Figure 4.1   Ogston in the Boer War. MS-3850-1-6-00097 in the University of Aberdeen Museums and Special Collections, licensed under CC By 4.0.

While the queen's sanction may have smoothed Ogston's relations with the British military, this did not, apparently, extend to the army medical authorities in South Africa. And Ogston recorded: 'a sense of humiliation in saying that such impediments as were placed in my way came from the head of the army medical department there'.[19]

Ogston sailed to the Cape aboard the *Mexican*. Among his fellow passengers were Sir Francis Grenfell, the Governor of Malta; 'three gallant public-school men' who had 'brought their horses along with them, and hoped to get commissions in some of the irregular mounted corps'; additionally, there were assorted volunteers 'of all ages from nineteen upwards, engineers, artisans, and even officers [. . .] all burning to take a part'.[20] At Madeira, grim news reached the ship: General Buller had suffered a reverse while attempting to cross the Tugela River. Buller was to be superseded by Lord Roberts as Commander-in-Chief of the Forces in South Africa, while Horatio Kitchener had been made Roberts's Chief of Staff. The *Mexican* docked at Cape Town on 4 January, and, after checking into the Mount Nelson Hotel, Ogston visited the Attorney-General for Cape Colony, Richard Solomon. (Solomon was a cousin of Bella's: his mother, Jessie Matthews, was a sister of Bella's father, James Matthews.) Solomon's position in South Africa was difficult. As Ogston summarised things: 'He had become a professed Dutch sympathiser, & joined the Dutch party's ministry – so that all he did was suspected and misrepresented in the newspapers.' Ogston, however, believed Solomon was 'a loyal & conscientious man, who had unwisely got himself into a very difficult position & was striving to do what he believed to be loyal and right'.[21]

Ogston had intended to join Buller's forces at Natal, but now decided to await the arrival of General Roberts; General Forestier-Walker, the GOC (General Officer Commanding) in Cape Colony, whom Ogston visited, also advised him to avoid joining Buller. Reading the *Cape Times*, Ogston was cheered to learn Sir William Stokes – his companion during their student days on the continent – was coming out to be a Consultant to the Army at Cape Town.[22] While he waited for Lord Roberts, Ogston gauged the state of affairs in the region. The majority of the Dutch population in Cape Colony were in the main, Ogston concluded, loyal to Britain and indisposed to support a Boer uprising. In terms of the military situation, there were four main lines from which the British could attack the Boer forces, heading inland, respectively, from the ports of Durban, East London, Port Elizabeth, and Cape Town via railway toward the inland Boer positions.

Ogston decided the Durban route was too remote to reconnoitre in the time available and, instead, visited the other areas while he awaited

Roberts. Ogston journeyed to the most westerly line first, running from Cape Town to Kimberley, where Lord Methuen's troops were now stationed by the Modder River. Methuen (1845–1932) was later to be General Officer Commanding-in-Chief in South Africa (1907–12), promoted to Field Marshal in 1911, and made Governor of Malta (1915–19). Forces under his command had been dispatched to relieve the siege of Kimberley but were blocked by entrenched Boer forces at Magersfontein on 11 December 1899. Methuen's setback was – alongside General Gatacre's defeat at Stormberg and General Buller's failure to relieve Ladysmith – one of the three battles constituting the 'Black Week' of December 1899 which prompted Lord Roberts's supersession of Buller.

At Modder Camp, Ogston presented his letter from Arthur Bigge and was introduced to Methuen, who in turn presented him to his staff, including the principal medical officer, Colonel (later General Sir Edmond) Townsend. Inspecting the typhoid hospital – which was located in a partially built schoolhouse – Ogston found it was over-crowded, no charts were used, and the invalids were attended to by Royal Army Medical Corps orderlies rather than nurses. The ambulance service, meanwhile, consisted of an assortment of carts and buck-wagons, which Ogston described as 'the roughest means of transport for wounded one could possibly conceive'. He added: 'Even the eyes of the non-medical officers could not fail to perceive that things were not as they ought to have been.'[23]

Ogston returned to Cape Town on 12 January 1900, where he obtained permission to visit General Gatacre's forces at Sterkstroom. The following day, Ogston boarded the *Dunnottar Castle* steamer where he was introduced, by the Earl of Errol, to General Kelly Kenny, who was travelling to Port Elizabeth to take up command of a force at Naauwpoort. On board, Ogston was wearing the khaki uniform he had procured in Cape Town, as a result of finding his civilian clothing rendered him conspicuous in the company of military figures. A high-ranking army medical officer in Kelly's force took exception to Ogston wearing this and, as Ogston recounted, 'wrote I believe to headquarters complaining of my wearing the Queen's uniform without authorisation'. The complaint caused Ogston no difficulty, but was 'an instance of the attitude adopted towards me by some officers in the medical services due to the part I had taken in advocating reforms'.[24] Ogston reached East London on 16 January. Here, the habitually heavy seas necessitated a novel mode of disembarkation, as Ogston and his fellow passengers were lowered, five at a time, in a large bottle-shaped basket from the *Dunnottar Castle* into

Figure 4.2   The armoured train General Gatacre lent to Ogston. MS-3850-2-1-B2-00001 in the University of Aberdeen Museums and Special Collections, licensed under CC By 4.0.

Figure 4.3   Ogston's Basuto pony and buggy. MS-3850-1-7-00139 in the University of Aberdeen Museums and Special Collections, licensed under CC By 4.0.

a tugboat. Ogston landed 'with a bump' and 'tumbled out' to hear the words, 'Oh, it's only a soldier. It won't hurt him.' Ogston was, in fact, unscathed, though he remarked the escapade 'set my teeth dancing, and was ignominious'.[25]

Ogston made his way to General Gatacre's camp at Sterkstroom on the morning of 17 January. Born at Herbertshire Castle, Stirlingshire, Gatacre had commanded a division under Kitchener at the Battle of Omdurman in September 1898 and had been sent to South Africa as Commander of the 3rd Division. Despite his forces being repelled during a night attack on the Boer positions at Stormberg, Ogston was impressed with the general, who he found living in 'a state of the most Spartan simplicity' in a railway carriage. Especially gratifying was Gatacre's offer 'to place at my disposal an armoured train in which I could proceed to the outlying parts of the country'.[26] Gatacre also invited Ogston to a lunch of Bovril, stew, and rice pudding, which was followed by port and cigars. By the time the latter was finished, Ogston recorded with understated, though perceptible, relish, 'I was told that my armoured train was in readiness, and that the General had placed orders that it was to be put at my disposal to go wherever I chose'.[27] Ogston took the train to Gatacre's advanced positions opposite the Boer forces at Stormberg, then returned to Sterkstroom, where – dining again with the general – he paid Gatacre 'the compliment of putting on my last pair of clean cuffs'.[28]

Ogston next wished to join the line held by General French's troops to the west. Rather than return to the coast and journey inland from Port Elizabeth, Ogston decided to travel cross-country through enemy territory. To obtain transport, Ogston travelled from Sterkstroom to Queenstown, where he was pleased to report:

> A local paper of the day contained I was informed, an announcement that 'Professor Ogston had arrived in the Camp at Sterkstroom – he is a handsome man of fifty-six, and he does not show any frills'. My best thanks to you, Mr. Editor! I am told the last phrase – which is either camp English or Colonial English – means – does not give himself airs! Merci![29]

Wearing a Red Cross brassard, Ogston travelled to Cradock. Here, he found there was a train which could take him as far as Naauwpoort, but from there to General French's position at Rensburg there was no certain means of transport available. Deciding to proceed to Naauwpoort in any case, Ogston telegraphed the PMO there, asking for a space to spend the night. Unbeknownst to Ogston, however, the PMO at Naauwpoort was Colonel Gubbins, who had taken such exception to Ogston's wearing uniform aboard the *Dunnottar Castle*. Gubbins ignored the request, and Ogston was preparing to spend the night on the platform of Naauwpoort station, when a kindly lieutenant of the Royal Engineers insisted Ogston take his bed for the night. Next morning, Ogston caught the train to Rensburg. General French was absent,

but Ogston met with the PMO, Lieutenant Colonel Donovan, and sur-
veyed the medical arrangements. Ogston found the medical staff lacked
serums and a proper operating table, only coarse dressings had been
provided, and that French's column was being serviced by improvised
railway ambulance carriages (an ambulance train having been denied
on account of the expense).

Having visited French's positions, Ogston returned to Cape Town on
23 January. Following Roberts's arrival at the Cape, Ogston visited him
in person and was advised to delay setting off for a fortnight until the
area of the campaign in which 'there was most to be seen' could be
determined. Less cordial was his meeting with the PMO of the South
African campaign, W. D. Wilson. In his journal, Ogston described the
encounter:

> Next, I had an unpleasant interview with Surgeon General Wilson, the last
> that shall take place with my will. He seemed intentionally to keep me wait-
> ing a long time, when he was quite disengaged, and then when he saw me
> was frigidly polite. He asked if I wore uniform. I said yes, I had a right to do
> so equal to that of any officer in the Army. Did I use the distinctive badges
> of rank or branch of the service? No, I wore the ribbons of my medals & the
> Red Cross. Did he object to these? He did not and could not object to my
> wearing khaki dress and helmet, nor the Red Cross or medals.[30]

Ogston was cheered, however, to be reunited with Sir William Stokes,
who had likewise 'seen Surgeon General Wilson & found him frigid'
and who had 'told [Stokes] there was nothing for him to do'. Ogston
concluded: 'It is evident that in Main Barracks the officials of the RAMC
are jealous of if not hostile to the Civilian Surgeons.' Likewise, Ogston
found the RAMC unwilling to accept any aid from voluntary societies
such as the Red Cross and Good Hope Society.[31]

While at Cape Town, Ogston also visited David Gill (a fellow
Aberdonian and Astronomer Royal at the Cape) and lunched at Groote
Schuur, Cecil Rhodes's country house. On 27 January Stokes, whom
Ogston described as 'looking so white & thin that I was grieved he should
go there' departed for Natal.[32] Later that day, Ogston inspected the military hospital at Wynberg and found, though admirably staffed by army
and civilian surgeons (the latter being present because the RAMC did not
have sufficient surgical staff of its own), the hospital lacked electric saws
and drills and antitoxin serums. The wooden wards were so infested with
insects, moreover, that mosquito nets were employed to prevent bugs
from falling onto the patients below, while the legs of the patients' beds
rested in cans of kerosene.[33] The following day, Ogston was invited to

lunch with Sir Alfred Milner at Government House, and he sat conversing with the Governor until Horatio Kitchener was announced, whereupon Ogston finished his 'cigar in the aide-de-camp's room'.[34]

Ogston was conscious of being present at a moment of profound
historic importance. Cape Town at this juncture recalled, to Ogston,
accounts 'of the state of Brussels in the summer of 1815, just before the
Battle of Waterloo. The city simply swarmed with distinguished and representative individuals.'[35] Among these was Rudyard Kipling (a fellow
guest at the Mount Nelson Hotel), whom Ogston described as 'an ugly
man'.[36] Despite his proximity to several eminent personalities, Ogston
was growing impatient of remaining in Cape Town. After the allotted
fortnight, he applied to General Roberts to be 'attached to some part
of the forces where I could study the methods of bringing aid to the
wounded under the modern conditions of warfare'.[37] Much to his satisfaction, Ogston was attached to Lord Methuen's column at Modder
River – the most westerly line of attack, and the one Ogston anticipated
Roberts would advance first. As a matter of courtesy, Ogston called on
Surgeon-General Wilson to ask for his official sanction. Wilson reluc-
tantly granted this, and told Ogston 'it would be "no kid glove campaign," but would be full of hardships and fighting, and that he would
give me neither tent, service, nor transport for my baggage'. Ogston had
hoped to be equipped with these, but – as he 'was not a man of feather
beds' – proceeded, undeterred.[38] He left his watch and medals at the
head office of the Standard Bank and deposited his rifle, umbrella, and
dress clothes at Messrs A. R. Mackenzie and Co. (Ogston's decision
to travel without weapons may have been intended to obviate further
claims he was operating beyond his remit as a civilian.) A leather port-
manteau, kit bag, and Ulster coat constituted, for Ogston, 'all I shall
require in this phase of the campaign'.[39]

Ogston's initial investigations of the medical situation in South Africa
had not been reassuring. He found the Royal Army Medical Corps
intransigent in the face of criticism and reluctant to accept aid from
voluntary societies or civilian consultants. The Red Cross, meanwhile,
could only make a negligible contribution, owing to a lack of transportation which meant its services could not reach the front. These deficiencies were nothing, however, compared to the almost biblical scenes of
destitution Ogston would witness thereafter, where the shortcomings of
the RAMC were rendered graphically apparent by the degradation and
suffering experienced by sick and wounded troops.

Modder River Camp had swelled from an encampment of 12,000 to
some 60,000 troops: 'the whole plain on the south was a huge city of

canvas as far as the eye could reach, while far away on the surrounding eminences glittered the lamps of the outposts flashing their messages to the centre'.[40] Ogston called on Colonel Townsend and was attached to the First Divisional Field Hospital of Methuen's column. To his pleasure, Ogston found this to be under the command of Major Coutts, a former pupil. Despite having been told by Surgeon-General Wilson he would only receive food, Coutts procured Ogston a tent, and a mess-sergeant agreed to supply him with a bucket of water each morning. Thus, Ogston 'with a thankful spirit [. . .] laid a mackintosh on a stretcher for a bed' and considered himself 'better off than many a man on the same plain'.[41]

Located approximately a mile from Methuen's headquarters were the First Divisional Field Hospital's two enteric fever hospitals. Enteric fever (also known as typhoid) was spreading through the army with alarming virulence, and Ogston was dismayed to find the hundred or more typhoid patients at Modder were being attended to by only three nurses and 'a set of orderlies [. . .] unfitted to be in charge of the management of men so seriously ill'.[42] Aghast, Ogston observed flies 'swarming on the faces of the insensible men, swarming even inside their mouths, and then conveying the poison everywhere'. Fatalities were common – fatalities Ogston believed, in part, to be avoidable:

> One's heart grew full to think how many of these men might have been saved, but for the fact that our rich country was ready to spend its money on everything save in organising in peace time the proper care of the sick and wounded in war.[43]

In an act of literary allusion which came to define Great War literature, Wilfred Owen famously emphasised the incompatibility of modern conflict with heroic notions of warfare. 'Dulce et decorum est pro patria mori', a line from Horace's *Odes*, translates as 'How sweet and wonderful it is to die for one's country'. If, however, Owen's audience had witnessed trench warfare, they would not, the poem's narrator claims, 'tell with such high zest / To children ardent for some desperate glory, / The old Lie: Dulce et decorum est / Pro patria mori'.[44] Anticipating Owen by more than fifteen years, Ogston wrote: 'Surely Horace had not seen all the sides of warfare when he wrote of the soldier's life: 'Momento cita mors venit aut Victoria laeta' ['In a moment comes either death or joyful victory'].' Similarly, then, Ogston evoked classical literature to emphasise the reality of modern combat. He continued: 'had our rulers spent a few weeks' in the Modder and other Boer War hospitals, 'seen what they had to reveal, and compared them with our best civil fever hospitals [. . .] they would have carried an undying remorse with them to their graves'.[45]

Following General Roberts's arrival at Modder River on 9 February, the majority of troops stationed there were dispatched. Along with only a few thousand, Ogston remained in the town of now-empty canvas tents. Orders had been issued, however, that all personnel must sleep in their boots and be ready to depart imminently. On 15 February, cavalry troops under the command of Lieutenant-General Lord French relieved Kimberley, a diamond mining town which had been besieged by Boer forces for a total of 124 days. On the 17th, Ogston was informed he was to join a force heading to Kimberley. Having struck down their tents, however, a sandstorm of 'wind and blinding dust lasting for some hours' arrived. And, having lost approximately 200 wagons to the Boers, Roberts was obliged to requisition all remaining transport carts, even ambulances. Ogston and his comrades were, therefore, left waiting alongside heaps of luggage on the otherwise barren plain when 'the worst sandstorm we had yet seen came over us, and lasted all night, burying us under inches of dust'.[46] When ambulance wagons began arriving from Jacobsdal, hospital tents were unearthed and re-erected in order to house the incoming wounded.

As Ogston's journals reveal, his own health suffered during the campaign. On 23 February, he recorded: 'Was rather poorly during the night on Wednesday–Thursday with sickness & vomiting, but [. . .] the dysentery promised to subside.'[47] Seeing Ogston was indisposed, a kindly Dr Grieg offered to ride to his quarters and bring some Lazenby's soup. Grieg, however, was thrown from his horse and dislocated his shoulder. After some difficulty, for 'Grieg was a large & powerful man', Ogston managed to reduce the dislocation by the Hippocratic method of placing his heel in the axilla, thereby providing a lever with which to adduct the arm.[48]

Ogston, accompanied by Major Pallen, took the train north to Kimberley, where he hoped to meet Herbert Grierson's brother Bernard 'Barney' Grant Grierson (1875–1946), a volunteer in the Diamond Fields Horse. Pallen and Ogston called at Kimberley Sanitorium, the residence of Cecil Rhodes (a diamond magnate and former President of Cape Colony) who was at lunch. Along with his card, Ogston sent in a letter of introduction to Rhodes written by Major Laing (Ogston met Laing, who was appointed commander of Lord Roberts's bodyguard, on the voyage to South Africa). Ogston and Pallen were hopeful of being asked to join Rhodes for lunch. However, they 'waited in a drawing room nearly half an hour, no one appeared, & the footman did not take a very broad hint I gave him that we should like tea or some refreshment'. Indignant, Ogston left and sent a message to Rhodes that he was unable to wait any longer.[49] While his initial attempts to find Grierson were fruitless, a Captain O'Mara, who knew of Grierson's whereabouts, intervened. O'Mara telegraphed Grierson's camp, found he was at present stationed out on the veld, and requested Grierson be let off duty to meet Ogston at the Kimberley Club that evening.

Cecil Rhodes later dispatched a messenger to Ogston, requesting he dine with him that evening. But Ogston declined the invitation and went to wait for Grierson at the Kimberley Club. Here, Pallen and Ogston were 'magnificently received and told that everything there was at our disposal. Mr. Brice, an influential citizen, took us to the bar, where we were to order whatever we desired'. The magnificence, however, 'dwindled when we came to particulars. There was no whisky, no soda water, no ginger ale, no biscuits, no anything we asked for, save lime juice, water pumped from the mines, and tea. So tea we had and were grateful for it'.[50] Grierson, possibly because the message never reached him, was unable to make the rendezvous, and Ogston and Pallen were obliged to leave for the 7 p.m. train to Modder. The train, it transpired, was actually a series of trucks. The resourceful Pallen, however, told the station authorities that Ogston was 'an enormous swell', which resulted in a carriage being added to the front of the train for them to ride in.[51]

As the fighting moved away from Modder, the camp's military significance diminished. However, the 'brunt of dealing with the invalids of Roberts's army necessarily fell on Modder River', with the result that Modder became 'neither a field hospital, nor a base hospital, nor evacuating station, but [. . .] partook of the functions of all three'.[52] This situation exposed the inadequacies of the Army Medical Department. Patients, sometimes in convoys of several hundred, poured constantly into the understaffed and poorly equipped field hospital at the Modder River Camp. Continuous new arrivals meant soldiers had to be constantly evacuated by train, even those 'requiring only a few days of rest and treatment to enable them to resume duty'.[53] At one point, Ogston recorded, there were 'only medical officers to attend 223 patients'.[54] The sick were laid on the ground within tents, without blankets or means of washing.

New admittances were transported to the camp in ox wagons over long journeys on primitive roads. Most wagons lacked any covering, meaning invalids were exposed to the sun and rainstorms. Ogston remembered, 'all were filled with men who were sick, men who had dysentery, men with torn hands and limbs, men with fractured bones [. . .] men shot through the head, through the chest, through the hip, or through the shoulders and arms', and were brought in by oxen – oxen who often died of exhaustion the moment they stopped.[55] It 'almost moved one to tears', Ogston recalled, to witness the arrival of patients in this manner, 'for their clothes had become so foul from their own and their neighbours' discharges, since the convoys could not be stopped for their necessities, that they had nothing on but a blanket and a helmet'.[56]

The overcrowded camp absorbed constant new intakes of patients, who received treatment prior to being sent on to Orange River and Cape Town. As the volume of invalids expanded, additional medical staff were sent to Modder. Medical supplies, however, were scarce and erratically packaged. Ogston described the situation of a medical officer:

> coming hopefully to get something particularly needful, and having to wait about and open boxes of 50 to 80 pounds, some of which had only methylated spirits, some only turpentine, and some only the splints which would have been invaluable 10 days previously, but which he no longer required.

Meanwhile, Ogston encountered only one Red Cross representative, who – owing to having to use military transport – arrived after the intensity of the medical work had subsided, and whose utility was reduced to distributing pyjamas.[57] Furthermore, only one of the hospital marquees had beds. In the other three, patients lay on stretchers or beds improvised from valises. On Sunday, 25 February, Ogston dressed twenty-five patients, and 'no light work it was' for a tall, middle-aged man, 'to stoop [. . .] bending & kneeling' from 9 a.m. to 2 p.m.[58]

On 4 March, however, Ogston was pleasantly surprised 'when a young man limped up to me, & in his smiling face I recognised Charlie Ogston'.[59] This was Charles Ogston (1877–1944), the son of Ogston's cousin Alexander Milne Ogston of Ardoe. Charles was serving in South Africa as a lieutenant in the Gordon Highlanders. (During World War I, he would be mentioned in dispatches and win the Distinguished Service Order, later retiring as brigadier-general.) During the Battle of Paardeberg, Charles's ankle had been accidentally crushed by a Canadian soldier, and en route to Modder he endured a journey of three nights and two days in a rough bullock cart. Ogston found him, however, 'looking brave and well, a fine, manly, modest, fellow, & with his sunburnt face & ribboned tunic, every inch a soldier'.[60]

That evening, a thunderstorm approached. These were common occurrences and, having checked his tent ropes, Ogston went back to the marquee. There, alongside Major Coutts, he was watching the approaching lightning:

> when in the space of a few instants a flush of hissing blinding drenching rainwater came flying past the door. A squall of wind instantly followed it, tearing up all one side of the marquee which fluttered & thundered & flapped in the wind. And as we clung to the poles to prevent everything going, we looked through the open side at where the camp was, a few yards from us, but an occasional glimpse of a tent through a grey sheet of water & small hail was all we could discern. The cracking thunder & hissing rain, & flapping canvas combined into one common roar in which each could not be distinguished, & the lightning flashed every second or two in the now darkening night, & soon all was dark save for its constant illumination.[61]

One of the hospital marquees and over thirty tents collapsed and 'in their place were sodden flat masses of writhing patients covered with tent poles, wet canvas & nooses of tent ropes'.[62] Ogston joined in the frantic effort to free patients trapped under the collapsed sodden canvas.

Having fixed disordered dressings and distributed his remaining dry clothes, Ogston then turned in for the night and attempted to sleep:

> But it was in vain. At 12 o'clock I was awakened by a dash of thunder, rain against my tent, & found there was a storm passing straight over us. The lightning was so constant that it quivered with a continuous light, transparent through the wet canvas walls and the wind roared tossing the tent, & the thunder overhead made the ground shake & tremble. Then came a roar like the sea in full fury & my tent pole wavered & bent, & down it came, across what it is my humour to call my waist, pinning me to the ground on my bed, & immediately I found I was unable to move.
>
> The only thing was to lie still & think. I lay with the wet canvas over my face & felt pools of water run down into my ears, & down under my body, & up along my feet and legs, & the canvas flapped, the thunder shook everything, & the lightning every instant or less poured out its streams of light as if the last day were come. From the commotion I knew many other tents were down, & that in time "The Professor" would be missed and rescued, so I just lay still.[63]

He was eventually freed by Major Coutts and joined in the work of re-erecting the flattened canvas city. Ogston found Charlie, spirits undeterred, struggling out of a collapsed marquee, which 'lay like a newspaper on a road in a rainy day'.[64] Ogston invited him to share his tent and produced his last tin of Huntley & Palmers biscuits and a piece of chocolate given to him by Bella. There, joined by Coutts, the trio 'made what we all thought a luxurious meal'.[65] Preparing to turn in for the night, Charlie, as had become his custom, scraped a hollow in the sand in which to rest his hip. Ogston cautioned him: 'Be careful what you're doing, I've got an arm buried there.'[66]

Charlie left for Cape Town via ambulance train on 6 March, and four days later, the entire hospital decamped and trekked north across the veld to Kimberley. In place of the camp, which had housed 50,000 men, was 'a naked, desolate expanse which showed only the crosses of the dead in the two little cemeteries, with the names upon them already half obliterated by the weather'.[67] On foot, Ogston accompanied the mile-long train of ox wagons, which halted at 9 p.m. The men lay down on the stony ground for a few hours rest until 3 a.m., when the march resumed. Having halted at 5 a.m., the convoy waited for the heat of the day to subside. Ogston, sheltering under a buck wagon, was content to spend the afternoon reading Horace. On the second night, the column bivouacked at Wimbledon, prior to continuing to Kimberley, which they reached in time for breakfast. That night at the Kimberley Grand Hotel, Ogston hosted a dinner for the officers of the field hospital. He described the occasion:

> we had many nice things: Hock, Champagne, Curaçao, Chartreuse, turkey, soup, fish, good cigars, sweets, and so on, and revelled in the unwonted luxuries of clean napery serviettes and finger glasses.[68]

Building on this success, Ogston was also able to locate Barney Grierson, who was stationed in the camp of the Diamond Fields Horse. Barney, whom Ogston approvingly described as 'a great strong soldierly looking man', had been working at the Roodepoort Mines when he joined the Diamond Fields Horse in order to participate in the defence of Kimberley. When retrieving the body of one Colonel Scott-Turner from where he had fallen at Carter's Farm, Grierson's 'comrades had been shot beside him, & had bespattered him with their blood and brains, yet he himself was never even grazed'.[69] With Barney's assistance, Ogston obtained a small patrol tent for his own use from an army store. Barney dined with Ogston and shared his tent that night, and Ogston telegraphed Herbert Grierson and Mary to let them know their son was safe and well.

At Kimberley, Ogston and his fellow residents also enjoyed the luxury of uncontaminated water. They could drink, bathe, and brush their teeth without fear of typhoid poisoning. Three days after their arrival, how-ever, no sanitary arrangements had been made for the new camp:

> I cannot find words strong enough to convey on paper my surprise and dis-appointment of such a condition of things, of course the ground all around us is rapidly becoming a cesspool: and the inevitable result must follow, in spite of our having pure water conveyed by pipes from the town reservoir.[70]

Sure enough, Ogston suffered a second attack of dysentery in March. By the 23rd, he reported this was wearing off and that the camp was in constant expectation of orders to move to a new location. To carry him and his baggage on the upcoming journey, Ogston purchased a brown Basuto pony and a light single-horse Cape cart.

On the 29th orders were received to march to an unknown destination, although it was anticipated they would pass Boshof initially. On 2 April, camp was struck and the company proceeded to the Kimberley Public Gardens, where they rested for the night. Then, after marching across the plain in the early morning, the company halted at a farm to outspan and breakfast. Here Ogston's tent proved an asset. One side could be opened up, making a shelter from the sun – one Ogston's comrades were happy to share. Having reached Boshof on 4 April, Ogston was alarmed to wake the next morning and find his pony had slipped his halter. His attempts to relocate the 'truant' were fruitless. To Ogston's delight, however, the pony was found by some horse artillerymen searching for their own charges.

Ogston's field hospital was ordered to accompany an advance of three or four thousand men which was to proceed under the command of Lieutenant-General Douglas. Extreme rain and an overhead thunder-storm meant these plans were cancelled, however. Fierce driving rain washed over the ground and penetrated the canvas walls of the tents:

> As I lay in what I must call my bed, the rain poured down on my face, & even my helmet got soaked through & could not keep it off. I had perforce to lie and soak, for the ground was wetter than the bed. Presently the rain softened the ground where the tent pegs were driven in: they came out as if stuck in butter as the wind rose & the canvas & ropes shrunk, and for some hours I was a soaking object in stockingless boots, going alternately out & in, strug-gling with pegs, ropes, & canvas to prevent the whole tent coming down, by working in the dark groping for the pegs & getting a whack at them now and then when a lightning flash showed their position, or chancing a stroke when it did not do so.[71]

The next day, at the request of Captain Mason, commander of the hospital's stretcher bearer company, Ogston agreed to assist with the care of three men whose wounds were too severe for them to be transported back to camp. Mason and Ogston journeyed the twelve miles to Tweefontein Farm, site of the Battle of Driefontein, where British forces had recently surrounded 100 Boers and killed their general, de Villebois-Mareuil. The three men were encountered in a fit state to be moved and returned to Boshof. Later that day, Mason found Ogston and led him to witness General Villebois's funeral. Behind the hospital camp 'drawn up in two rows facing one another, were about ten thousand soldiers, each with his rifle resting vertically with the muzzle on the ground'. Along the pathway formed by the soldiers, Villebois's body was carried to the burial ground. As the afterglow lingered in the west, 'thunder clouds were again gathering & the twisted streams of the lightning were playing' while the 'last post' was sounded. Ogston described the scene with admiration as 'a neat ending to a valiant soldier's life'.[72]

The Boshof camp, Ogston wrote, was '[q]uite shut out from the world [. . .]. We have no newspapers letters, or telegrams, & form a little world of our own with its own interests, hopes, & fears'.[73] Relative inactivity and a lack of definite news led to the circulation of rumours. On Easter Sunday, Ogston recorded a list of 'Today's Camp Rumours', which included:

> That Lord Roberts has got 7,000 Boers surrounded by his forces near Bloemfontein.
> That Mr. Beck, the merchant near Market Square, has prophesied that we are about to have a great deal of rain.
> That the Hon. Geo. Gough, cashiered for his feeble command of the cavalry at Belmont, has committed suicide by shooting himself.

To these Ogston added a list of 'Camp Facts':

> That the cold weather has almost put a stop to the mosquitoes, to my great relief.
> That the whole camp resounded with loud and repeated cheering tonight when a ration of rum was ordered on account of the great cold & thunder showers, very bad this evening again.
> That the question of stockings is becoming a very serious one with me.[74]

On 23 April, Methuen, attempting to allay the Boshofians' fears the hospital was contaminating the local water supply, ordered the hospital be repositioned half a mile away. This new location – behind the new cemetery – Ogston complained, was 'not a good place – by any means', one which positioned 'us between the lines of defence & the outposts'.[75]

From this isolated position, it was difficult to obtain clear intelligence of the campaign's progress. However: 'One undeniable and mysterious set of facts remains, viz: - That Lord Methuen has, with 5 or 6 thousand, retreated before a force of some 2,000 men and is entrenching himself here in a way quite unlike what might have been presumed.'[76] Ogston's health, meanwhile, continued to worsen. On Sunday, 29 April, he recorded being 'too ill to write since Friday, from another attack of violent diarrhoea [. . .] I have never, save for a day or two at a time, felt quite well since the middle of February, have had two attacks of dysentery, pretty bad, & constant diarrhoea'.[77] So weak was Ogston that, overcome by a spell of faintness, had to hold on to the arm of Mr Newman, a volunteer surgeon, in order to steady himself. Keeping the true extent of his infirmity to himself, Ogston conceded his poor health might cause him to return home.

Temporarily feeling better, however, Ogston introduced croquet (he had a croquet lawn at Glendavan) as a means of occupying the hospital staff's spare moments. This was played on a small plot – the only grassy one to be found locally – using croquet balls borrowed from a Boer prisoner. Tent-peg mallets and hoops from car tyres completed the other equipment necessary for several successful games to be played. Ogston, in fact, began to find the camp's relative inactivity a trial:

> It is beginning to be deadly dull in Boshof Camp, and in the fine weather we now have, we resemble a set of lotus-eaters, we haven't even the happiness of possessing something to grumble at. Our friends the Boers are still around us, but will neither attack nor fight when we attack them.[78]

On 7 May Lord Methuen invited Ogston to dinner. Talk ranged from tents, to typhoid, water supplies, and munitions. The latter subject allowed Ogston to showcase his impressive memory, as he recorded in his diary: 'Curiously none knew how many balls or what size were in the Vickers-Maxim shells; and I was the only one who knew the weight of a shrapnel ball (11:13 grams, say 170:200 grains, or rather less than half-an-ounce).'[79] Having enjoyed a couple of Cuban cigars, his after-dinner tranquillity was disturbed upon exiting, when a sentry pointed a rifle muzzle in Ogston's face and demanded he produce the countersign. After uttering the correct word ('Jacobsdal'), Ogston was allowed to return to his tent.

Methuen's conversation at dinner indicated the column was soon to be in action, and Ogston was eager to participate. His wellbeing, however, continued to be a concern: 'Every morning [. . .] I have some hours of dull grinding abdominal pain in the right side (caecum or ileum) and would have constant diarrhoea did I not control it with opium and bismuth.'[80] Ogston suspected typhoid but was so intent on remaining with Methuen's forces that he did not take his temperature, in case this revealed him to be in grievous ill health. He was pleased, however, to receive indication his efforts were valued by Methuen's troops, when Captain Mason:

> told me that I was now recognised by everyone, from the General down-wards, as the Consultant of this force. I have never requested any position at all, but have quietly worked away to help in my special sphere, without advertising myself, and it gives me pleasure to know that this opinion has been formed of me & my work without solicitation or obtruding myself.[81]

Ogston's precarious health held out until 15 May, when the order to advance was given (the destination was unknown but generally believed to be Hoopstad). The column was to proceed surrounded by cavalry at 6 a.m., then wait out the heat of the day until 4 p.m., when they would continue until 7 or 8 p.m.

The column got underway at noon, with each unit carrying twenty-one days' supplies. At times on the trek, Ogston was visibly fatigued. However, by the kindness of Major Coutts, Dr Newman, and Captain Mason, he was given occasional extra rations or help mending his harness or pitching his tent. Passing a farm located next to a small dam, Ogston followed other riders in driving his thirsty pony to the water. The pony, however, sank into the muddy ground, became entangled in the – now stuck – buggy, and began to sink. Ogston leapt into the mud and gradually extricated the buggy and panicked animal. Having reached Aaronskraal, word was received from Methuen that Hoopstad had surrendered without resistance.

One morning on the trek, Ogston was making use of a quiet moment to dig up some bulbs from the plain, when he encountered a group of horsemen chasing a wounded bull, which then approached Ogston. Mistakenly believing it to be an ox, Ogston – who happened to be carrying a hammer – calmly prepared to defend himself. The bull, however, then charged at an unfortunate soldier 'engaged in an occupation which must not be told', who made an undignified scramble behind a mound of earth.[82] Upon learning the animal in question – which was chased out of sight – was a bull, Ogston 'thought it well I didn't have to depend too much on the hammer'.[83]

The next phase of the route to Hoopstad was a waterless fifteen-mile stretch which was tackled in a brutal forced march. From sheer exhaustion, oxen and mules collapsed in their harnesses and were left on the veld to either expire or recover. 'It was a hell of cruelty,' Ogston remarked, to witness the oxen being driven on by the ceaseless lashing of their drivers, or left – their necks raw from the harness abrasions – lying on the ground over which the convoy passed.[84] Hoopstad – a town of 150 inhabitants, by Ogston's reckoning – was reached on Sunday, 20 May. Sick and wounded soon poured in, however, so that by evening there were 110 invalids to care for. The landdrost's (magistrate's) hall and the schoolhouse were commandeered, and beds and bedding were requisitioned from local hotels and citizens.

The next day, the column began its trek to Bothaville. The terrain to be crossed was less severe than that leading to Hoopstad, but the nights were cold and Ogston, though somewhat better, was still hampered with diarrhoea. 'I keep it in check with opium,' he recorded, but he consciously had to avoid overtaxing himself.[85] Ogston was also suffering from rheumatism and had developed a swelling in his right knee. Bivouacked near a loop of the Vaal River at Zandfontein, Ogston commissioned one of the attendants to cut him some hay to make a more comfortable bed. Upon returning, however, Ogston discovered three ponies trampling his tent and eating the hay. Although his telescope was flattened, Ogston was consoled to find his camera had survived.

On Saturday, 26 May, the column set off from Bothaville in the direction of Kroonstad. Ogston's health, however, was declining rapidly. His diarrhoea worsened, which he continued to treat with opium pills. Upon reaching their next stop at Nieuwejaarspruit, Ogston was too ill to do anything more than lie down beside his cart. Ogston's pony and tent were attended to by his comrades, however, causing him to 'hope to live to remember them with gratitude for many years'.[86] Again, after the column reached a subsequent halting place called Doornspruit, Ogston was so fatigued: 'I could do nothing but celebrate the occasion by omitting to wash my grimy hands & face, and also enhanced the distinction by leaving the stumps of a dirty beard. It must have been amusing to see me.'[87]

Kroonstad – already engorged with sick and wounded – was reached on Tuesday, 29 May. In addition to a canvas RAMC field hospital, two churches and a hotel had been converted into hospitals. Moreover, there were no fresh supplies, meaning sick and well alike had to subsist on tinned goods and biscuits. Ogston himself was by this point an invalid. He conceded feeling 'not very clear in the head', and Major Coutts:

> said if I were one of the officers or men of the force, he would make me go sick & give it up. And I inclined a favourable if somewhat stupid ear to his counsel & told him he should dispose of me as he would.

I was getting no better, had been living for 2 months or more with nearly continual diarrhoea, not kept in check by repeated daily doses of opium, Dover's Powder, Bismuth, Chlorodyne, & a fine array of astringents to aid them in their duty. I was useless for any work, if military work had come, I could not have gone to see it or helped in it: for I was too weak to walk a hundred yards, & I was wasted till my limbs were contemptible & my stomach had become a deep hollow.[88]

Moreover, it was clear that the Boer forces in the Transvaal were fading and unlikely to mount even a last stand against the British. Coutts – to whom Ogston gifted his horse and buggy – told him he 'had been of the greatest use to his hospital & the column under Methuen, that it would take the pleasure from the rest of the campaign for him that we should no longer be together'.[89]

Ogston was put aboard a south-bound ambulance train which departed Kroonstadt at 11 p.m. on Wednesday, 30 May. He began to feel better, but was shocked, when undressing to bathe, to witness in the mirror:

the spectre of a gaunt old man whose long grey hairs, eyebrows, & moustaches overhung a face the size of a child's, whose great limbs showed only snakes of empty muscle like dead flaccid eels hanging from his skeleton like the cordage from the masts and yards of a ship, I hurried into my clothing again thinking of the poor horses who had been left on the veld to their fate.[90]

At Bloemfontein, Ogston transferred into a train heading for Cape Town, and slept on board as he waited for the morning departure. When he woke the next morning, however, Ogston knew immediately he was seriously ill. He was then transferred to Bloemfontein's Upper Dames Institute and treated for typhoid. Ogston summarised the severity of his situation:

I had been going about duty for 25 days with the fever upon me and at least eight days with phlebitis in the veins of my right leg, & on admission my temperature was 104°, and such circumstances usually involve an attack of the greatest danger, with probable intestinal perforation, or haemorrhage.[91]

Ogston lay delirious for days. His description of this state of detachment and psychosis, told in characteristically precise prose, makes for what is perhaps Ogston's most remarkable literary achievement:

I lay and slept near the door of the 4 bedded ward they placed me in, a constant stupor excluding any fears or thoughts. I knew the huddled mass lying like a tumbled chair near the door was me, yet it was not I. For I used regularly to get up with something dark & soft in my left hand, I know not what, and wander away under grey sunless moonless starless skies, wandering and wandering to the horizon, seeing other shades glide silently by and solitary but not unhappy, till something stirred the mass by the door. Then I was drawn back to it, & it became I, and was fed, & spoken to, and cared for. And when they left it again, I wandered off as before by the side of a great slowly flowing flood, through my fields of asphodel knowing no difference between daylight & darkness with no thought of religion and no fear of the death I know was near, and roamed through thick grey skies apathetic and content till they disturbed my body lying near the door and I was again drawn back to it and entered it with something like disgust.

Ogston recalled being well enough to give two feeble handclaps on 5 June to celebrate the news that Lord Roberts's forces had taken Pretoria, but the state of intense delirium persisted until the 7th, after which his 'wanderings seemed to be fewer & shorter, & the mass at the door and I grew together and ceased to be separate and never again ceased to be one and the same'.[92]

Ogston was full of praise for the medical staff at the Upper Dames Institute and noted in particular his appreciation for being 'fed on champagne and sedulously nursed by Sister Warrender and Canadian Sister MacDonald'.[93] With the chief danger over, Ogston was moved to a two-room ward where he convalesced over the coming fortnight. His strength returning, Ogston was able to write letters to Bella, Lord Methuen, Colonel Townsend and other acquaintances from the war – but it took him until Friday, 22 June to catch up on his journal entries. Excepting the absence of a headache, Ogston had suffered a severe case of typical typhoid – one complicated by phlebitis of the veins in his right leg. He could very easily have died, as did five of the nursing sisters and several of the army medical doctors who caught the disease at Bloemfontein.

As Ogston's health recovered, he progressed from 'arrowroot, egg flip, jelly, oatmeal porridge & finally bread and milk, a climax of triumph reached on the 25th June'. Then, on the 27th, he smoked his first cigar since admittance ('ah! how sweet it tasted!'). On the 28th, Ogston was placed on a stretcher and moved to the veranda. 'The delight of lying baking in the sun again, & seeing trees & sky around, was a sensation to be remembered, after so long a confinement,' he noted, continuing:

> But how sadly the glaring blue African sky wants a friendly cloud or two. Ah! Give me the mist clouds trailing along the hillsides, the snow clouds wrapping themselves round the mountain summits [. . .]. Well – please God and I may yet see them all again, & shortly, and be surrounded with those I love at home.[94]

Ogston seemed to be aware his journals would be read by his family – certainly by his wife, partly for whose benefit he kept accounts of his adventures. Rarely, however, did he record his personal life with any manner of emotive effusion. This passage indicates Ogston was fully aware he had come close to death – to an extent resulting in such longing for the topography and domestic life he enjoyed in Scotland.

Despite his admiration for its staff, Ogston was unimpressed with the hospital as a whole. Few of the hospital's orderlies were RAMC men, the rest being untrained regimental soldiers. Furthermore, although the hospital could accommodate fifty-three patients, it did not practise disinfection in anything like an appropriate manner. The sole apparatus for disinfection was 'one enamelled hand basin containing a creolin solution' which resided on the veranda. Meanwhile, wards 'saturated with enteric poison' were merely swept, not disinfected, with the sweepings being discarded near the veranda. And no antiseptic soap was used unless a patient happened to possess some. Consequently:

> on leaving this, patients covered with the enteric poison are put in the trains, a hospital train it may be, but quite as often the saloon carriages of an ordinary mail train, and are sent down country poisoning the trains, and disseminating the virus until they reach some hotel, private house, or steamer, where they can obtain baths and so gradually purify themselves while shedding off the poison into these places.[95]

It was no surprise, Ogston commented, medical staff often contracted virulent forms of typhoid.

By Wednesday, 4 July, Ogston was well enough to travel by rail to Cape Town. He had been promised a compartment of his own, but this, Ogston discovered on arrival, was already occupied. The Railway Staff Officer, however, attached an extra carriage to the train, so Ogston could have a compartment. This may have been in deference to Ogston's status as an eminent royal surgeon, but he was clearly still in a fragile state. At the station, two different officers noticed Ogston's faltering perambulations and gave him their arm to lean on. On board, Ogston was cheered to open a box of provisions (given to him by one of the nurses at the hospital) and find it contained 'biscuits, six raw eggs, two bottles of Calves-foot jelly, loaf bread, a piece of cheese I had petitioned for, a bottle of milk, and one each of champagne and port, and a knife and teaspoon, and some of the food and stimulants soon revived me'.[96]

At Naauwpoort, Ogston was transferred to a sleeping compartment in a saloon corridor carriage which ran to Cape Town. Here he was invited to stay with the Hon. Mr Justice William Henry Solomon and his wife, Maude. Justice Solomon (the brother of Richard Solomon, Attorney-General for Cape Colony) was a further cousin of Bella's, whom Ogston had stayed with during his 1894 trip to South Africa. Ogston attended to some practical exigencies, such as cabling Bella to confirm his safety, booking passage home on the *Briton*, and obtaining a haircut and new shoes 'so as to appear less wild & savage'.[97] He was also delighted to be reunited with William Stokes, recalling with uncharacteristic levity: 'Sir William Stokes has grown a beard!!! Dear old Stokes!'[98] The Solomons' residence, Ballyrush House, had a ground floor parlour where Ogston recuperated, looking out to the gardens, smoking, reading, and writing as he desired. His hosts were, if anything, too forthcoming in their solicitation:

> Special things are cooked for me, especially fish, which is good for me, which I like, and which is very good here; & puddings: various pretexts are found to make me take champagne, which I am too ready to do. I must, when I leave thence, lead a hard Spartan existence as to food & drink, for I am become a gluttonous man & a wine-bibber among the kind people here.

And such generosities were on hand at every hour, for Ogston slept next to a night stand laden with port, whisky, biscuits, and cheese. He concluded: 'The temptations I am subjected to are worse than those of Saint Anthony.'[99]

Among Ogston's fellow passengers on the *Briton*, which departed Cape Town on 11 July, was Lady Sarah Wilson (1865–1929). Wilson, an aunt of Winston Churchill, became one of the first female war correspondents when she was asked to cover the Siege of Mafeking for the *Daily Mail* in 1899. Ogston recorded, with some disdain: 'Lady Sarah Wilson, of Mafeking fame, smokes cigarettes [. . .] The cigarette sticks in the corner of her mouth, removed constantly to permit a high falsetto voice to utter small talk and badinage. Though she looks clever, I have not heard a single clever or sensible thing said by her.'[100] Ogston was even less enamoured with the Duke of Marlborough 'a fair-faced pleasant looking non-entity' and Duke of Norfolk 'a little bilious skinned, straggling black-bearded creature who looks as if he had liver disease & felt it'.[101]

Of the remaining notables, only Arthur Conan Doyle gained Ogston's approval. Doyle, who had practised as an ophthalmologist before concentrating on his literary career, had been serving with a hospital sent to South Africa by the philanthropist John Langman. According to a RMS *Briton* concert programme dated 18 July 1900, which Ogston preserved, Doyle favoured his fellow passengers with a reading from his 1896 novel, *Rodney Stone*. Ogston noted in his journal: 'to hear Conan Doyle read from his own work "Rodney Stone" the scene of the race for the bridge was a thing to be recalled in after time. A fine manly modest great-hearted fellow'.[102] On 23 July, there was a fancy-dress ball, at which Conan Doyle appeared in a 'very good' costume as Chang the Chinese Giant. Ogston, however, 'was not in the frame of mind to enjoy' such levity. 'The faces of the dead men I have known since last I passed here kept coming up before me, and I went early to bed.'[103]

Also on board was a French military attaché named Duval – presumably Roger Raoul-Duval (1877–1917) – who had been accompanying the Boer forces. On Friday the 20th, Ogston recorded an incident where Duval – '[a] pert forward little slip of a Frenchman':

> had been airing some of his views that the British used expanding bullets in this war, until he was taken to task, banished from the table where he previously sat, & had to unsay his words to the officers he had offended.

Duval's mistake, as Ogston saw it, emanated from the fact 'that during the voyages out in the transport-ships the Mark IV (Woolwich) expanding bullet was served out for rifle practice'. However:

> The Mark IVs were all recalled before the troops landed, but some of the revolver ammunition was carried ashore & retained by our officers for a month until the solid bullets were sent out & issued in its place. But this mistake had no practical result, for the revolvers were never fired.[104]

Each side accused the other of firing Dum-dum bullets during the war.[105] Belief the British forces employed such ammunition may be attributed to the fact that Boer forces had found some of the, ostensibly unused, Mark IV rounds. Meanwhile, it is possible the discovery of Mauser hunting ammunition in Boer positions prompted British supposition the Boers were firing such rounds at them. Again – aligning with Ogston's view that the wounding potential of contemporary ammunition was imperfectly understood – claims of Dum-dum use may also have been brought about by 'misjudging the effects of modern, high-power rounds and the effects of multiple hits with repeating rifles'.[106]

The evidence Ogston gathered during the war – anecdotal and first hand – indicated, however, that the Boers were using expanding bullets. He heard claims at Belmont that Mauser big-game rounds had been found on Boer troops.[107] At Modder River Camp, a medical officer reported finding '[m]any expanding bullets', including 'a soft nosed Mauser', in a captured Boer entrenchment.[108] And, again at Modder, Ogston recorded being told by Major Molesworth of the RAMC that he 'had seen several explosive wounds and large quantities of captured expanding bullets'.[109] Meanwhile, while inspecting a captured Boer position at Boshof, Ogston determined: 'Expanding bullets had been freely used by them, here lay one dropped in haste, there a posy of them arranged for convenience in rapid firing eleven in number, and all these expanding slitted, Dum-dum pointed bullets.'[110]

Rather than casually record these examples as they arose, Ogston was keen to investigate the matter further, and he made repeated requests to be apprised of any evidence of expanding bullets being used by the Boers. At Modder River, for example, he 'asked Major Coutts to allow his officers to show me any of the wounds they might meet with which might have the appearance of having been produced by expanding bullets'.[111] The subject of expanding bullets was discussed during his dinner with Lord Methuen and his staff, and Ogston recorded they 'promised to give me samples of all the different kinds of expanding bullets they could'.[112] During his convalescence at Bloemfontein, meanwhile, Ogston sketched under the heading 'Expanding Bullets' some Boer ammunition he had acquired, including a 7.0 mm soft-nosed expanding Mauser bullet captured at Brandfort and a 0.8 mm 'man stopping' Willes Mauser Pistol bullet. He also sketched a bottle-shaped 0.577 Express bullet 'used against us and found at Bloemfontein', which Ogston was given.[113] These and other examples he recorded indicate his interest in the issue of expanding bullets continued throughout the war. Possibly because the evidence Ogston gathered was insufficient, however, he does not appear to have made any public claim the Boer forces employed exploding bullets during the war.

As the *Briton* neared home, Ogston reviewed his state of health. He had recovered almost all his lost weight, but bowel issues and a swelling in his right leg persisted. Ogston's sight and hearing were also compromised, and he envisaged being unable to participate in grouse shooting that autumn. After an uneventful passage, the *Briton* crossed the English Channel, and Ogston was met at the wharf by Bella. He reached Aberdeen on 29 July 1900, having been absent seven months and fifteen days. In a series of supplementary notes added to his journal, Ogston noted that William Stokes died of pneumonia in Natal on 18 August 1900. Ogston described him as 'My dearest friend' – to which he added a quote from Tennyson: '"The sweetest soul that ever looked with human eyes" God rest him!'[114] Ogston did not append a record of the death of his son, Hargrave, however. According to the inscription on a memorial to Hargrave in St Nicholas Church, Aberdeen, he had been serving as a trooper in the Imperial Light Horse and was killed in action at Riet Kuil, Klerksdorp, Transvaal, on 17 April 1901.[115]

5

1901–1913

*[H]is reputation is worldwide, and in every part of the British Empire there
are old pupils of his who look back to their student life with gratitude to
him for his teaching and example.*[1]

Following his return from the Boer War, Ogston undertook a
series of new honorary and advisory roles. In February 1901, he was
named a supernumerary Surgeon Lieutenant in the Aberdeen Company
of the Volunteer Medical Staff Corps – the organisation he helped establish upon returning from the Soudan in 1885.[2] Then, following Queen
Victoria's death in January 1901, Ogston was appointed (Honorary)
Surgeon in Scotland to Edward VII (he would later attend the coronation of Edward and Queen Alexandra at Westminster Abbey in October
1902). Next, on 13 June 1901, Ogston received the degree of LLD from
Glasgow University.

As Ogston had anticipated, and seen confirmed by personal experience, the Boer War emphasised the need for further improvements in the
recently established RAMC. Chief among the medical concerns vivified
by the war was the unacceptably high number of soldiers incapacitated
by disease. During the conflict, the RAMC treated some 22,000 soldiers for wounds and injuries, while 'twenty times that number were
admitted to hospital with disease; 74,000 suffered from enteric [fever]
and dysentery alone'.[3] Moreover, the capacity of the RAMC was rapidly
overwhelmed by the growing scale of the campaign, necessitating the
enrolment of civilian surgeons.

In June 1901, Ogston was made a member of the Expert Committee of
the Secretary of State for War to discuss the reorganisation of the RAMC.
The *BMJ* noted Ogston was the only civilian member of the Committee
who had taken interest in military medicine prior to the war, and that –

despite his incendiary remarks at the 1899 BMA gathering – Ogston's 'practical experience during two campaigns' and 'intimate knowledge of the army medical systems of Germany and France' meant 'his views are certain to carry great weight with the Committee'.[4] Ogston's expertise was called upon again in October 1902, when he was asked to give evidence before the Royal Commission on the state of the army regarding 'the equipment and organisation of the Army Medical Corps compared with those of Germany and Russia'.[5]

Summarised in the *BMJ*, the 'Report of the Royal Commission on The South African War' (1903) included Ogston's views that, with reference to medical orderlies in the war, many of these were 'absolutely ignorant of anything like what was required for attending on the sick'. Meanwhile, army medical surgeons, urged Ogston – reiterating his observations at the 1899 BMA conference – required greater experience of modern surgery than they could acquire while simply practising within the army during peacetime:

> Our operations nowadays are pieces of very high art which a man acquires by daily training, weekly training. [. . .] [I]f a man has not that daily practice he may know the theory most perfectly, he may be a most able and intelligent man, and yet he will not do such good technical work as the, perhaps, less able man who has had this practice.[6]

Meanwhile, the 'quality and quantity' of equipment supplied to the forces in South Africa 'were defective, and generally they were antiquated and badly organised', and 'the whole system of drug supply had not been adequately thought out'.[7]

In light of such criticisms, the RAMC underwent significant redevelopment. Its Medical School was relocated from Netley to London, where it became the Royal Army Medical College; the Army Nursing Service was replaced by the Queen Alexandra's Imperial Military Nursing Service; and a School of Sanitation was established at Aldershot in 1906. Furthermore, to gather the guidance of eminent civilian medical practitioners, an Army Military Advisory Board was established, to which Ogston gave evidence on contagious diseases in March 1904.[8] In December, meanwhile, he accepted an invitation to become an Examiner in Surgery for the RAMC in London.

In 1902 Ogston also published further work on the treatment of club foot, this time in relation to young children. Rather than surgically remove tarsal bones, in part or whole, Ogston advocated a different approach. Since the ossification of tarsal bones in children is not fully complete, Ogston proposed using a curette to remove the cancellous centre of the tarsal bones which opposed rectification of the deformity. 'A tarsal bone which has in this manner been deprived of its osseous centre is soft and plastic [. . .] and very moderate pressure upon it suffices to mould it into any required shape'. The interior of the repositioned bone would then fill with clotted blood which would be 'as in a simple fracture, gradually replaced by cartilage which slowly ossifies; and the bone, with all its ligaments and joints, is intact'.[9]

March 1902 saw Ogston appointed to the Royal Commission on Physical Training for State-aided Schools in Scotland. He was one of nine members of the commission, which was chaired by the Earl of Mansfield and sat twenty-eight times to gather evidence from 127 witnesses.[10] The report concluded that 'education cannot be based on sound principles which neglects [*sic*] the training and development of the bodily powers'. To bring this into effect, it was recommended there be larger provision of playgrounds and exercise halls in elementary schools; that 'games and physical exercises', including instruction in hygiene and physiology, 'should be treated as an essential part of the school course' in secondary schools; and that universities consider 'rearrangement of courses' to avoid overburdening students and leave opportunity for physical exercise.[11] Ogston joined the League's Executive Council in April 1905.

In November 1905, meanwhile, he was elected president of the Aberdeen Medico-Chirurgical Society. The society was co-founded in 1789 by James McGrigor and eleven other medical students, who sought to obviate the lack of medical teaching offered by both King's College and Marischal College. It may have been of some cheer to Ogston to reflect not only had his contributions to the establishment of the RAMC furthered McGrigor's work in strengthening the Army Medical Corps, but that he was now president of the society his great Aberdonian medical forebear had helped instigate.

And further honours, advisory positions, and senior leadership roles followed. In July 1907, Ogston became Medical Referee under the Workmen's Compensation Act 1906 for the Sheriffdom of Aberdeen, Kincardine, and Banff. Whereas the 1897 Act – to which Ogston had also served as a medical referee – was mainly targeted at industrial workers, the 1907 Act extended employers' liability to pay damages to all injured workers. Meanwhile, in November 1907, he was made Vice-President of the National League for Physical Education and Improvement. Then, in February 1908, at Christ Church College, Oxford, Ogston delivered a lecture on 'The Future of the Medical Profession' for the Oxford

Medical Society. He was made an Honorary Colonel of the RAMC's Territorial Division in April and, in July, elected Vice-President of the Aberdeenshire branch of the Red Cross Society.

While Ogston participated as a medical referee for workmen's compensation, his daughter Helen was engaged in another aspect of social reform: women's suffrage. Helen (1882–1973) was born at 252 Union Street and graduated with a BSc from Aberdeen University in 1906, whereupon, according to Elizabeth Crawford, she moved to London and qualified as a sanitary inspector. Helen joined the Women's Social and Political Union in 1908 and was an active speaker in the organisation, appearing alongside Marie Brackenbury and F. E. M. Macaulay at the Hyde Park demonstration in June 1908.[12] In December 1908, Helen attended a meeting of the Women's Liberal Federation at the Albert Hall, at which the main speaker was David Lloyd George – then Chancellor of the Exchequer for the Liberal government. Helen, along with other WSPU members in the audience, planned to heckle Lloyd George in protest of his superficial endorsement of female suffrage. In anticipation of violent response from the Liberal stewards, Helen equipped herself with a dog whip, despite Sylvia Pankhurst's recommendation an umbrella would form a more judicious means of self-defence.

As the WSPU members stood up and began protesting during Lloyd George's speech, Helen – who was in a second-tier box – was met with the following response:

> a man put the lighted end of his cigar on my wrist; another struck me in the chest. The stewards rushed into the box and knocked me down. I said I would walk out quietly, but I would not submit to their handling. They all struck at me. I could not endure it. I do not think we should submit to such violence. It is not a question of being thrown out; we are set up on and beaten.[13]

Helen's subsequent use of the whip to repel her assailants made the front page of the *Illustrated London News*, which ran the headline 'The woman with the whip: The Militant Suffragettes' new weapon in use at the Albert Hall' above an image of Helen about to bring an airborne whip down upon a quartet of male stewards.[14] Despite Pankhurst's reservations, she concluded 'several of the newspapers protested strongly against the behaviour of the stewards at the meeting', while the *Manchester Guardian* and *Globe* expressed sympathy with the WSPU's impatience at Lloyd George's evasive promise to introduce a reform bill which would not, under certain conditions, oppose female suffrage.[15] What did Ogston make of Helen's actions? Whatever his assessment, both father and daughter exhibited an indomitable willingness to challenge authority in order to secure reform. Walter Ogston described his father as 'a supporter of women's franchise'.[16] Helen, meanwhile, recalled her father's sympathetic support: 'Later, when many people were inclined to laugh at my suffrage outbursts, he always understood that it was my "David's pebbles being thrown at the Goliath of social injustice" that hurt me so dreadfully at the time.'[17]

In February 1909, Ogston's four-year term as a RAMC examiner expired, and in October he retired as Regius Professor of Surgery at Aberdeen University. In recognition of his exceptional career, Ogston was awarded a LLD degree by Aberdeen University in March 1910, while in October a portrait of Ogston – painted by George Fiddes Watt – was presented to the university and hung in the Mariscal College Picture Library. Meanwhile, in 1911, Aberdeen University instituted the Alexander Ogston Prize in Surgery, which provided a gift of surgical instruments to the best student in each year's surgery class. Ogston did refuse some accolades, however. When invited in November 1910 to stand as a parliamentary candidate for Glasgow and Aberdeen universities, he declined. Likewise, Ogston was unable to accept the Colonial Office's request he serve as British Representative to the 1911 Australian Medical Congress in Sydney. Moreover, although he retained his title as Honorary Surgeon in Scotland upon the ascension of George V in 1910, he was also unable to attend the coronation of George and Queen Mary at Westminster Abbey in June 1911. There is no evidence to suppose Ogston would have been suited by nature or inclination to become a politician, but his decision to miss George V's coronation and the Australian Medical Congress may indicate, upon entering his late sixties, Ogston was less disposed to undertake unessential travel.

Sometime after his purchase of Glendavan in 1888, Ogston developed an interest in the prehistoric structural remains extant in the Howe of Cromar, a topographical basin of land surrounded by hills – chiefly Morven (to the west) Pressendye (to the north), and Craiglich (to the east) – to which the areas around Loch Kinord and Davan belong. Ogston conducted significant field research to map and investigate these remains, and he wrote up his findings into a manuscript which was published posthumously as *The Prehistoric Antiquities of the Howe of Cromar* by the Third Spalding Club in 1931. According to the volume's editor, W. Douglas Simpson, Ogston appears, barring a few minor addendums, to have largely concluded this work by 1911.[18] In his Preface, Ogston outlined his belief that the remains found in the Howe constituted, in part, domestic dwellings dating from the Stone and Bronze ages.[19]

He carried out a series of detailed surveys, measuring distances by tape line, paces, and prismatic compass. To compare the prehistoric remains of Cromar with other ancient structures, Ogston visited numerous sites in Scotland (including the Hebrides), England, Ireland, Normandy, and Brittany.

Ogston wrote that, upon his return from the Boer War, Queen Victoria had intimated a wish to see him, but this 'fell through' due to her 'failing health'. He was assured by Sir James Reid, however, that Victoria 'had not forgotten' Ogston and had 'expressed the intention of conferring some honour upon' him.[20] Walter, however, claimed Ogston 'had refused' the offer of a royal honour 'more than once'.[21] Whatever the cause of this apparent reticence, in June 1912, Ogston was gazetted a Knight Commander of the Royal Victorian Order (KCVO), entitling him to a prenominal 'Sir' and for his wife to be known as 'Lady' Ogston. In October 1912, Ogston was treated for appendicitis by John Marnoch, his successor in the Chair of Surgery at Aberdeen University.

In February 1913, Bella died, having, according to Walter, 'been failing for some years'. Walter summarised their marriage: '[Bella] was not, intellectually, a companion to him, though they were devoted to each other and had a happy married life. This was especially so after her children had grown up and she was able to enjoy more freedom.' After the death of his second wife, Ogston lived with Flora, who was 'a semi-invalid'. 'They were, I believe,' Walter wrote, 'very happy together and I know that he was so'.[22] In July, crowning his ascent to the top of his profession, Ogston was named President-Elect of the BMA. Confirming Ogston's appointment, the *BMJ* noted 'his reputation is worldwide, and in every part of the British Empire there are old pupils of his who look back to their student life with gratitude to him for his teaching and example'.[23]

The events in Ogston's life between the Boer War and World War I – honorary degrees, honorary royal positions, advisory roles, a knighthood, and the BMA presidency – confirm his place as a revered, respected, and celebrated figure in British medicine. They also betoken the inevitable transition from active work towards retirement. At the onset of the Great War in 1914, however, Ogston – despite being in his seventies – still felt capable of being actively useful to his profession. And his service during 1914–18 would prove he did not overestimate his abilities.

6

1914–1918

I was not sorry to have the opportunity of seeing what modern war was like. It was all very different from my former experiences.[1]

A S WELL AS CONFIRMING Ogston's election as the BMA's next president, the 5 July 1913 edition of the *BMJ* informed readers the Association's 1914 annual meeting would take place in Aberdeen.[2] From January 1914, the *BMJ* featured a series of essays overviewing the history of the city, its principal buildings, notable inhabitants, and university. The *BMJ* also provided detailed information for delegates wishing to explore north-east Scotland and the Highlands, including the notice: 'Sir Alexander Ogston, is making arrangements to facilitate visits to places of archaeological interest by members who are interested in such matters [. . .] such as the Druidical circles of Kingcausie [and] the prehistoric fort on the Mither Tap of Bennachie.'[3] The same edition gave notice that Ogston's former pupils and dressers were organising a lunch in his honour on 30 July at Aberdeen's Grand Hotel.[4] The detail of planning in the *BMJ* – which even included a list of local golf courses and their respective merits – indicates, as late as summer 1914, the BMA planned a convivial, edifying meeting in Aberdeen, with every exertion being made for delegates to expand their attendance into a holiday.[5]

The event opened with the Annual Representative Meeting on 24 July, and, on the 28th, Ogston was officially installed as President for 1914–15 in a ceremony at Marischal College's Mitchell Hall. Later that day at the Music Hall, Ogston gave as his President's Address an account of Bishop Elphinstone's founding of Aberdeen University in 1495 and the early teaching of medicine there.[6] In the audience were Sir James Barr; Lord Provost of Aberdeen; Professor Sir T. Clifford Allbutt; the Right Honourable Robert Farquharson, MD, of Finzean; Ogston's daughters

Constance and Mary (and her husband, Professor Herbert Grierson); Professor Thomas Nicol, Moderator of the Church of Scotland; Bishop Chisholm; and Professor John Marnoch, Ogston's successor to the Chair of Surgery at Aberdeen.[7]

After Ogston's Address, Marnoch informed the audience:

> when the profession in Aberdeen resolved to invite the British Medical Association to meet there and the invitation was accepted, they looked in one direction, and one only, for a President. Sir Alexander Ogston was very reluctant and diffident, but such was their pertinacity and such their unanimity that ultimately he consented to lay aside his hobbies, to come out of retirement, and to lead them.[8]

That same day, the heir to the Austrian throne, Archduke Franz Ferdinand, was assassinated, triggering a series of declarations of war among the European powers.

Meanwhile, the lunch to honour Ogston on 30 July was attended by some eighty guests and presided over by Sir James Porter.[9] At the annual dinner later that day, Sir William Osler proposed a toast to Ogston, remarking the medical profession honoured him:

> first, for his splendid record as a surgeon – he was known all over the world in that capacity; secondly, as a teacher – a great and good one whose students were all over the world, carrying with them the best traditions of Scottish surgery; and thirdly, for the splendid example he had given of how to mature. No man in the profession had matured more gracefully or satisfactorily.

Osler (1849–1919) – a founding member of Johns Hopkins Hospital and creator of the medical residency programme for speciality training of physicians – was known for his sense of humour. He remarked that, since the death of the surgeon Douglas Argyll Robertson (1837–1909), 'my dear old friend of forty-two years has taken his place' as the 'Adonis of the profession'. Ogston, professing he did not possess the oratory powers to adequately reply to this compliment, said '[h]e would confine himself to the adequate yet inadequate word "Thanks"'.[10]

As such pleasantries unfolded, momentum was gathering that would soon precipitate global war. Arthur Anderson Martin, a surgeon from New Zealand – who would go on to win renown for his service with the RAMC during World War I – formed part of the international delegation in Aberdeen. He recalled:

> a very curious incident towards the end of the meeting – the last day of July. The president of the Association, Sir Alexander Ogston, gave a reception to all the delegates from the British kindred and affiliated associations, and

to the foreign representatives. Although the German and Austrian delegates had been about in the morning, not one was present at the evening reception. They had all departed silently, and had said goodbye to no one.

Germany and Austria had sent out their messages, and the medicals returned with all speed.

We were then on the eve of war, but none of us at Aberdeen thought that we would be in it, or that we were then swiftly rushing to great events.[11]

The 8 August edition of the *British Medical Journal* juxtaposed the success of the Aberdeen meeting with the unfolding gravity of the international situation:

The meeting at Aberdeen is now but a memory, yet one so pleasurable and bright that it must always serve as a striking background to corresponding recollection of the dramatic events which have since occurred and are still occurring. Never has a curtain at a theatre fallen and risen between scenes presenting a greater contrast than those at Aberdeen on Saturday and Sunday respectively. On Saturday the weather was bright and sunshiny [. . .] on Sunday there were torrents of rain, the streets were thronged with people bidding farewell to hundreds of fisher folk summoned to rejoin the navy, and many of the local officers and their visitors were deep in study of mobilisation.[12]

On 1 August, Germany announced war on Russia, with Britain declaring war on Germany three days later.

Prince Albert (the future George VI) was at this time serving as a midshipman on HMS *Collingwood* at Scapa Flow in the Orkney Islands. Suffering from appendicitis, the prince – accompanied by Sir James Reid – sailed to Aberdeen, where he was operated on by Marnoch. It seems Reid felt bound by courtesy to inform Ogston his successor would carry out the procedure. On 3 August, Ogston replied: 'I assure you I quite understand, and moreover approved. I am quite on the shelf now, and it is best so. Marnoch is the younger and better man, and the one who ought to attend the Prince.'[13] At Reid's suggestion, however, Ogston attended the operation at Aberdeen Nursing Home on 9 September. George V wrote to Reid: 'Certainly, I of course approve of Sir Alexander Ogston being present at the operation as you propose, his opinion and advice will [be] very useful.'[14] Alongside Marnoch and Reid, Ogston was a signatory on a bulletin to Buckingham Palace assuring the royal household the operation had gone well.[15] Albert returned to HMS *Collingwood* and later took part in the Battle of Jutland in 1916.

Despite having described himself as being 'on the shelf' – and being in his seventieth year – Ogston showed a characteristic determination to serve in the unfolding conflict. His initial offer of service to the Army Medical Department of the War Office was unsuccessful, he suspected, on account of his age. However, Ogston eventually secured a post as Honorary Consulting Surgeon at Southall Military Hospital from November 1914 to March 1915. Located in west London, the hospital was one of the largest and best equipped facilities run by the Voluntary Aid Detachment. The *BMJ* reported Ogston had relocated to the area and was in daily attendance.[16] James A. Davidson, a former student of Ogston's who served as an assistant commandant at the hospital, records Ogston being 'busily engaged in operating in a fine building, built to the design and under the supervision of his son Axel'.[17] This indicates Axel – Alexander Lockhart Ogston, Royal Marines – played a role in converting the building from its former purpose (a recreation hall for the employees of Danish firm Messrs. Otto Monsted & Co.) to a hospital. The hospital's Commandant, Dr E. A. Chill, described Ogston as 'a great surgeon and a great gentleman', possessing 'the gift of winning the esteem and affection of all with whom he came in contact'.[18]

Ogston was then asked to take charge of a detachment for the British Eastern Auxiliary Hospital on the Danube in Belgrade. Admiral Sir Ernest Troubridge, head of the British Naval Mission to Serbia, had requested a hospital for his men. This was refused on the grounds the naval surgeon already provided for his men was sufficient. However, sufficient funds for such an amenity were raised by Una Vincenzo, Lady Troubridge, better-known for her relationship with Marguerite Radclyffe Hall, author of the novel *The Well of Loneliness* (1928), a bestselling exploration of lesbianism. On 1 April – again joined by Flora and Davidson – Ogston sailed from Liverpool aboard the ss *Saidieh*. In his journal, Ogston transcribed a note from Clive Wigram, the King's Assistant Private Secretary and Equerry, which conveyed the encouraging message:

> Sir James Reid has told me that you and your Daughter are going to Belgrade to a hospital for British Seamen and Soldiers, which information I have communicated to the King. His Majesty hopes that you will have a good voyage and is sure that you will find lots to do when you get there [sic].[19]

Ogston recorded that, as the vessel passed the Scilly Isles and crossed the Bay of Biscay, seasickness – rather than the anticipated submarine attack – was the chief cause of anxiety aboard. The threat, however, was very real: the *Saidieh* was torpedoed and sunk in the mouth of the Thames by a German submarine on 1 July 1915.

Ogston landed at Salonika in Greece on 15 April and journeyed north via railway to Belgrade. He was impressed with the Serbian capital 'looking proudly down on the flooded Danube at its feet'.[20] Ogston described the city as containing 'a good many ruined houses' and 'partly destroyed shops and buildings' but, overall, as surprisingly undamaged despite its recent liberation from Austrian forces. The hospital was housed in the Third Belgrade Gymnasium, which had 'the appearance of a palace'.[21] This imposing palatial building was surrounded by acres of grounds, while, inside, a vast marble hall gave access to a series of 'well-lit rooms fifteen feet high, twenty feet broad, and from twenty to fifty feet long'.[22] The building, which had been used as a barracks by the Austrians, was cleaned and readied for receiving patients. The scale of the Gymnasium belied the level of medical activity taking place within, however. As Ogston noted, the military events in 'Belgrade at this time were of minor magnitude' and there were, at that point, 'only fifty British naval artillerymen in the country'.[23]

Flora had remained in Salonika while Ogston journeyed to Belgrade. On 25 April, he received news she had been ill for nine days with abdominal pains. He travelled south immediately and 'found her out of immediate danger'.[24] Ogston faced a dilemma. If Flora could be transferred to Belgrade, he felt, with the assistance of the staff there and his own '(perhaps diminished) operating skill, the chances would be better'.[25] On the other hand, there was the risk of leaving Flora 'a convalescent invalid' in a warzone. Furthermore, while the 'conditions of travel [were] not very bad', there was still the possibility Flora might require an operation 'at short notice' en route to Belgrade.[26] Despite the countering risks of bad weather or submarine attack, Ogston decided it was best to return to Britain.

Ogston described himself as caught between competing loyalties – to Flora on the one hand, and, the other, his profession in wartime. He was glad, though, to be 'able to give her confidence and comfort, and repay somewhat of the cares she had devoted herself to giving me'.[27] Additionally, Ogston questioned whether his, possibly waning, powers were equal to the task in hand: 'In my life before I have never been so near to feeling that the burden is greater than I can bear'. He wrote: 'the occurrence of a tragedy is or may be dependent on my soundness of judgement or steadiness of nerves in the next few days, and that I am alone in Salonika, with none who can advise me, that at my age my judgement may not be what it was.'[28]

On 10 May, Ogston and Flora sailed to Malta. Here, while they waited for the P&O steamer *Novara* to set sail, Ogston was thrilled to have the opportunity to visit the ancient megalithic temple complex of Hagar Qim. Alongside the stone circle at Callanish on the Scottish Isle of Lewis, Ogston deemed Hagar Qim 'the most interesting spot in the world'. He wrote: 'I studied it carefully, and took measurements, bearings, and photographs, to supplement the plans and views of it which I have at home.'[29] Evidently, Ogston's passion and knowledge of prehistorical remains extended far beyond those of Scotland. And Ogston was gratified to become reacquainted with Lord Methuen, now Governor of Malta. Lunching with his former general, Ogston found Methuen was 'taking a great interest also in the antiquities of the Islands'.[30]

On 27 May, Ogston and Flora reached London. Sir James Reid put Ogston in touch with Sir Alfred Pearce Gould, who operated on Flora two days later. Ogston's decision to return Flora to London proved astute. He described the operation performed on her as 'a bad one, difficult & prolonged by complications'. He estimated: 'she may recover, probably will, but the future for her is gloomy'. This experience clearly shocked Ogston, who wrote: 'I am an old man, & the last month of anxiety has turned my moustache grey; but I ought to be thankful, and I am, that at the end of a long life [. . .] I have been able to be of such real use to dear Flora.' Keen to return to Serbia, though disappointed to do so alone, he wrote: 'I care not a great deal if it be ordained that I have my thread of life cut off there.'[31]

No doubt the distress of possibly losing his daughter and companion coloured these remarks. However, both James Davidson and Ogston's son Walter asserted that, at least in part, Ogston hoped to meet his end in the 1914-18 war.[32] As his decision to retire early from the role of Senior Surgeon indicates, Ogston had the foresight to acknowledge even his abilities would wane. Nevertheless, as his subsequent war record shows, there is little to indicate his age was an impediment to Ogston providing useful service under frontline conditions. Moreover, he lived for another fourteen years – in which he pursued his numerous hobbies. Perhaps, following the recent death of his second wife – and now facing the loss of his daughter – Ogston did not relish the possibility of a prolonged senescence.

On 3 June, Ogston found his mood lifted by the 'marvellous recovery' of Flora.[33] He may also have been heartened by the news – printed in the 8 May edition of the *BMJ* – that the BMA Council recommended Ogston be re-elected president for 1915–16.[34] Possibly these positive developments were some consolation as he endured a farcical return journey to Serbia. Ogston's plan was to journey through Italy to Brindisi, where he would catch a steamer to Greece. In Paris he was advised to take the train to Milan, in order to take a further train to Italy. This, however, meant travelling through neutral Switzerland:

> the result was that when I stepped out of the train at the Swiss Frontier Station, I was, being in uniform, arrested, and told I was liable to be 'interned'

as being in uniform and carrying arms, of which I confessed being in possession of a sword and pistol.[35]

Once the Swiss authorities had established Ogston's identity, he was allowed to continue, although he remained in his sleeping compartment for fear of arousing further suspicion.

Worse followed in Italy, which had entered the war on the side of the Entente on 23 May. As he crossed Italy by rail, Ogston was – upon surfacing to arrange meals for himself and Sister Stokes, the nurse travelling with him – repeatedly suspected of being a spy. In Brindisi – where Ogston and Stokes were detained at length by the Chief of Police – the boat he had hoped to catch was cancelled. Next day, Ogston and Stokes took a train to Gallipoli on the Italian coast. Here, the next boat he had intended to meet was, he learned, stopping at Crotone instead. A thirteen-hour journey then followed 'round the instep of Italy', where there were 'fresh exhibitions of the spy mania from staring, jostling crowds, most of whom were inclined to be rude'.[36] The *Epiros*, which they had planned to board there had, however, departed. 'It was atrocious,' Ogston admitted, 'but there was nothing to be done but to bear the blow in silence.'[37] Ogston and Stokes found accommodation in the Hotel Pitagora. Next morning, there was still no boat. To kill time, Ogston attempted to explore the town, 'but the same spy mania was everywhere exceedingly unpleasant', and he returned to the hotel.[38]

The situation was growing desperate. Ogston was running out of funds, and the heat, harassment, bad food, flies, and perpetual disappointment were beginning to take their toll on one usually so unflappable:

> My brain is not yet clear enough to do justice to the change from yesterday to today. I don't think the mental balance will be restored for several days.
>
> When I wrote the preceding lines and was employing my siesta after lunch in Crotone in scraping the swarming houseflies off my fingers and pen, despairing of getting out of Italy to such a degree that I felt that it might all end in an apoplexy or mental derangement, or some tragedy. We seemed, when thrice successively the steamers had failed us, to be enmeshed in a web from which escape was hopeless. And Stokes, though as good a creature as lives, almost drove me crazy with her silly ways and silly empty remarks. (I hope I didn't show it.)[39]

Moments later, however, news reached Ogston that the *Mykale* had arrived. He and Stokes successfully boarded, and such was Ogston's good humour he cheerfully tolerated the mosquito that bit his fingers and bunged up his right eye at 4 a.m. 'I was still in a berth, still out of the nightmare of Italy, could still see with my left eye, & the creature was welcome.'[40] Finally, Ogston was able to sail to Greece, on to Serbia, and rejoin his hospital by 21 June.

Ogston's exertions to rejoin the Belgrade hospital were not rewarded, however. Upon arrival, he learned the hospital administrator had been recalled, meaning there was no one to oversee the hospital's finance, including payment of staff salaries. Just three days later, Ogston wrote: 'The game is played out, and tomorrow ought to see me take farewell of Belgrade for ever.'[41] Ogston was offered command of the hospital by Admiral Troubridge, but refused, partly to avoid the disparaging treatment the previous administrator had been afforded.[42] The hospital, in any case, had few British occupants in its wards; accordingly, 'there seemed to be no prospect of its fulfilling the functions for which it had been created'.[43]

After four months' involvement with the Belgrade hospital, Ogston returned to Scotland in July 1915 (during which month he was unanimously re-elected BMA President) and spent the next year occupied with numerous domestic wartime roles. At its meeting held in London on 23 June 1915, the BMA's War Emergency Committee of the Metropolitan Counties decided to extend their efforts to the whole country, thus forming the national War Emergency Committee – of which Ogston was an *ex officio* member.[44] He noted that, as part of the Committee, he responded to the request from Director-General of the War Office, Sir Alfred Keogh, for 2,500 more doctors. In this capacity Ogston was also a signatory on a letter to the press urging the public to remain loyal to GPs who were serving in the armed forces, arguing that it was a patriotic duty to ensure these doctors' domestic practices were preserved.[45] In September 1915, Ogston was re-elected President of the BMA's Aberdeen branch for the third successive year, while, on 9 October, he addressed the Aberdeen County branch of the Red Cross at Inverurie, lamenting the Geneva Convention was no longer being universally upheld in the present war.[46]

Ogston also attempted to organise the senior and retired members of the medical community for service in warfare. This he described as 'an – at any rate partial – failure', owing to resistance from the officialdom of the Army Medical Department and 'the jealousy of the Central Medical War Committee'.[47] More successful, in Ogston's estimation, was an article he published in the August 1916 edition of *The National Review* on the use of sphagnum moss as a field dressing. Ogston estimated that, over the next six months '45,000,000, or it may be something like 100,000,000, of dressings will have to be provided'.[48] Given the scarcity of gauze and cotton wool, Ogston encouraged greater efforts be made to employ sphagnum moss. Charles Walker Cathcart, Senior Surgeon to the Royal Infirmary of Edinburgh, had already published advocating its use in the present war. While the moss – which grows widely in Britain – was already being gathered (on the Balmoral estate, for instance) and converted into dressings, Ogston called for increased exertions to be made in anticipation of the vast scale of dressings that would likely be required:

> It is to be hoped that, in view of its probable extreme value, there will be no delay, during the favourable months, in collecting very large stores of the Sphagnum Moss, organising its preparation into dressings, and giving it a fair trial, on a large scale, as a material for treating wounds received on the field.[49]

As the armistice was more than two years away at this point, Ogston's call proved prescient. Despite such contributions – not to mention duties with the Aberdeenshire Territorial Association, the City War Work Association, the County War Work Association, and the Aberdeen Sphagnum Moss Joint Committee – Ogston keenly regretted not having a more practical role to play in the war. The seventy-two-year-old wrote: 'I was beginning to feel that the world had no more need of me.'[50]

Figure 6.1 The Villa Trento. MS-3850-1-7-00139 in the University of Aberdeen Museums and Special Collections, licensed under CC By 4.0.

The ill-fated Gallipoli campaign ended in ignominious withdrawal in January 1916. In June the British navy secured an inconclusive victory in the Battle of Jutland. Meanwhile, the notorious Somme campaign claimed tens of thousands of British lives. For someone with a keen sense of patriotic duty, a long-standing commitment to military medicine, and a zest for adventure, it must have been difficult to observe the course of the war from a position of, comparative, inactivity. At this time, however, Ogston was approached by the First British Ambulance Unit for Italy, who urgently required a surgeon and looked to Ogston for a recommendation. Given the scarcity of surgeons he proposed himself, and from September 1916 served as a surgeon with the First British Ambulance Unit in north-east Italy at the Villa Trento, located approximately ten miles to the west of Gorizia, where it maintained an ambulance base.

Leaving on 2 September, Ogston travelled to Udine where he was met by Dr G. S. Brock, who had left his position as Physician to the British Embassy in Rome to serve as the unit's Chief Medical Officer. Brock was in charge of medical matters, while the unit as a whole was superintended by the historian G. M. Trevelyan. Ogston and Brock waited for Trevelyan, who was travelling from Rome, to join them, before motoring to the hospital together. The Villa – leased by the Italian government from the Count of Trento – was 'a huge rambling building, with many annexes and outbuildings, facing the south' – situated a ten-minute walk from the foothills of the Alps 'amid gardens and grounds laid out in Italian style, with many beautiful and rare trees'.[51] The unit contained '3 sisters, 14 V.A.D. nurses, and (at least) 4 male dressers'. Patients were accommodated in two wards within the villa, while a large granary had been converted into two further wards. The yard buildings, meanwhile, served as a garage for the 'some 16 or 20 ambulances & motors' attached to the unit. The routine, as Ogston recorded, was 'called 6.30: breakfast 7.00: lunch 12: tea 3.30: dinner 7.30'.[52] On Ogston's arrival there were several 'bad' surgical cases but no new ones; however, on his second day Ogston was able to 'save an arm (I hope) from amputation by resetting the elbow joint shattered by a bad gunshot wound'.[53]

At the villa, Ogston formed part of a remarkable concentration of intellectual and creative figures. The members of the unit were individuals who due to age, 'medical rejection, Quakerism, and conscientious objections' could not serve in the army.[54] The Friends' Ambulance Unit – mainly composed of young Quakers – had been operating in Ypres since October 1914. Among the Friends were Philip Noel-Baker (1889–1982) – who would go on to be an Olympic silver medallist, Labour MP, and Nobel Peace Prize winner (1959) – and the noted mountaineer

Geoffrey Young (1876–1958). In May 1915, Young, Baker and other Friends were added to the staff of Unit One of the British Red Cross in Italy. While unit members operated as part of the Red Cross and wore khaki, it was independently financed by the British Committee in Aid of the Italian Wounded.

Trevelyan, whom Ogston described as 'a thin spare, anxious man, myopic with spectacles, & a long, stooping, hatchet faced man', was a former fellow at Trinity College, Cambridge, chiefly known for his trilogy of works on Giuseppe Garibaldi.[55] The son of Sir George Otto Trevelyan, 2nd Baronet, Trevelyan left Cambridge in 1903 to focus on writing but later returned Cambridge, in 1927, as Regius Professor of History and later became Master of Trinity College. Ogston esteemed Trevelyan's abilities, and the impression was reciprocated. Trevelyan wrote: 'we also had the honour to have with us Sir Alexander Ogston, whose fame and whose quiet, benevolent courtesy to all persons great or small added to the prestige and popularity of Villa Trento'.[56] 'In the ever-recurring seasons of disappointed hopes and unexpected disasters,' Trevelyan wrote, 'nothing did me more good than [. . .] to see [. . .] Sir Alexander Ogston smile as he smoked his pipe.'[57]

Freya Stark (1893–1993), who would go on to have a notable career as a travel writer of the Middle East, worked as a VAD in the villa's Garibaldi Ward. Stark described the strains of working at the hospital in her autobiography *Traveller's Prelude*:

> We washed in a little tin basin, dressed in ten minutes or so; made our beds (mattresses filled with straw); breakfasted at 7.30 and worked from 8, as far as I can remember. There was a bath, and one put one's name down for it on a list, but someone from the outstations always pinched it. I think I only got one in two months. [. . .]
>
> This was the first time I helped out at an amputation; it is a strange and shocking thing to feel a limb become suddenly lifeless in one's hands. I think I saw why nurses are nearly always happy people: their life is constant drama, with no interval of boredom – people are always recovering or dying. With all the carnage that was going on, to be helping to save life was a comfort.[58]

Ogston, a long-time advocate for the benefits of female medical workers in wartime, noted the villa's 'women officials, sisters, and nurses were a well-trained, well-educated, well-behaved body [. . .] whose conduct entitled one to be proud of them'.[59]

Thomas Ashby (1874–1931) – a classical scholar, archaeologist, and Director of the British School at Rome – served at the villa as a translator and store master. Ashby's antiquarian expertise endeared him to Ogston, who appreciatively remarked Ashby 'especially interests me as being a consummate archaeologist, who has the subject of early Mediterranean civilisation at his fingertips'.[60] Henry Tonks (1862–1937) teacher (and later professor from 1918–1930) at Slade School of Art – who had previously been a house surgeon at London Hospital – served briefly as an assistant surgeon at the villa, before going on to accompany John Singer Sargent on tours of the Western Front as an official war artist. Also present were Edward Garnett, who acted as D. H. Lawrence's editor for *Sons and Lovers* (1913) and the painter Elliott Seabrooke (1886–1950), a former pupil of Tonks's at the Slade Institute. So esteemed were the Villa Trento staff, the King of Italy took an interest in the hospital and even paid 'a surprise visit to the Villa Trento' to confer on Trevelyan the Italian Silver Medal for Military Valour.[61]

In addition to the Villa Trento hospital, Unit One maintained two ambulance stations at Gorizia under the command of Geoffrey Young. And Trevelyan's was not the only Red Cross unit sent out to Italy. Unit Two, based at Tolmezzo, was superintended by the sculptor Francis William Sargant (1870–1960). Unit Three was stationed at San Valentino. Unit Four comprised a mobile radiographic facility operated by Countess Helena Gleichen (1873–1947) and Nina Hollings (1862–1948). Gleichen – a landscape painter and relative of Queen Victoria's – was the daughter of Prince Victor of Hohenlohe-Langenburg. Hollings, meanwhile, was the sister of composer and suffragette Dame Ethel Smyth (who also visited the Villa Trento). Together, Hollings and Gleichen enlisted in the Red Cross, first serving as ambulance drivers in France. They then learned radiography during a six-month course of study in Paris and raised funds to finance their own unit. The French, after accepting their offer, merely attempted to appropriate the unit's equipment for themselves. Hollings and Gleichen went on to serve with great distinction in Italy and were both awarded the Bronze Medal for Military Valour.[62] Unit Five, meanwhile – a further radiographic service – was headquartered at Verona under the command of Cecil Pinsent (1884–1963), a noted landscaper who remodelled several Tuscan villa gardens in the style of the 1500s.

On 9 August, the Italians began to retake Gorizia. That night, while under constant bombardment, Geoffrey Young made over twenty trips across the bridge into Gorizia. Ogston remarked Gorizia was 'the only city of capital importance which had been taken from the enemy in the whole war'.[63] Shortly after his arrival, Ogston was driven out to Gorizia to survey the damage. The city, surrounded by Austrian-held San Marco, San Daniele, and San Gabriele, was under constant bombardment; in return, Ogston remarked, '[t]he whole city and its suburbs seemed to be vomiting cannon shot'.[64]

Having retaken Gorizia, the Italians were keen to capitalise on their gains and make further conquests of Austrian-held territory. The Austrians retained the mountains in the Julian Alps to the north of Gorizia and, to the south, the much-coveted city of Trieste which lay beyond the Carso plateau. In autumn 1916 'the whole county, from Udine to the Isonzo' was a scene of immense activity, as preparations were made for further attacks on the Austrian positions. During this time, Ogston estimated hearing 4,320 cannon shots from the Italian artillery per hour. As he recalled, the bombardment had spectacular results at night: 'the horizon on the crests of the hills was lit up by the sparkle of the exploding shells which threw up fountains of earth and rocks high up into the air; the more prolonged glitter of the star shells and the searchlights [was] white as an aurora borealis or tinted like the early streaks of a summer's dawn'.[65]

Depending on the volume of casualties from the front, the hospital served both as a clearing station and stationary hospital. It had initially been 'intended to receive patients for a day or two and send them back to Base', yet it was decided that in periods of relative quiet it could be kept as a stationary hospital 'retaining any patients whose conditions made it desirable that they should remain'.[66] Ogston writes: 'It was not uncommon for us [. . .] to have to evacuate the patients in our wards until perhaps only three unremovable cases remained, and in a couple of days later two or three hundred would be sent in, so that all of our beds were again filled.'[67]

In October 1916 – to his relief, during a lull in admittances – Ogston's health declined sharply:

> For ten days or so I have been troubled with Diarrhoea, kept in check by Chlorodyne. On the 4th, however, after ward visit was over, I had a furious attack of Dysenteric diarrhoea, intense pain almost to fainting, and was fortunate to meet Dr. Brock as I crept back to my room. He ordered me at once to bed, put me on milk alone, gave me Caster oil, followed by Bismuth and Laudanum, & shortly I felt better, but for 2 days I kept bed, & today only am up, though not yet quite able to do ward or other work.[68]

Ogston was soon able to return to his surgical duties, however, and he was able to offer further practical assistance to the unit. The ambulances used by Unit One were, after a year of continuous service, becoming worn out. Ogston wrote to William Smith – an Aberdeen advocate and head of the local Red Cross branch – asking him to put his appeal for a new ambulance before the Aberdeenshire Red Cross Society. Smith felt the matter too pressing to wait until the Executive Committee's next meeting, and so sent Ogston's letter to the *Aberdeen Daily Journal*, which printed it on 25 September – along with the news Smith and his wife had opened the subscription with a £10 donation.[69]

On 19 October, Ogston noted in his journal he was 'very proud of the <u>Response</u> Aberdeen has made' to his request for ambulance funds:

> Flora gave £5: Walter gave £5, and it is creeping up so as to have reached £80 or £90. Ardoe [Ogston's cousin] gave £5. Even if it is not sufficient to get a car, it will help towards one. But Mrs. Davidson, York House, Cullen, sent an offer of £500 to Trevelyan, and Lady Sempill diverted the car she had intended for Russia to us here.

Meanwhile, Smith had passed Ogston's appeal to their colleagues in Glasgow. Hector Cameron – formerly Lister's house surgeon, now Sir Hector Cameron, retired Professor of Clinical Surgery at Glasgow University and Red Cross Commissioner for the West of Scotland – wrote to Ogston promising four or five ambulances from the Glasgow Red Cross.[70]

Queen Elena of Italy (1873–1952) visited the hospital. Elena worked as a nurse during the war, transforming the Villa Margherita and Quirinal Palace into military hospitals. She raised funds for these endeavours by selling signed pictures of herself at charity events – thereby inventing the signed celebrity photograph. Ogston's journal notes she spoke to all but one patient (who was asleep) and took a great interest in everything mechanical in the hospital. Their conversation was carried out in French, as Ogston's Italian was 'too elementary to be useful in a conversation'.[71]

Later that month 'trouble came' when Trevelyan was informed members of the unit who were of military age would be withdrawn and conscripted.[72] Unit One arrived in Italy before the Derby Recruiting scheme came into effect in autumn 1915, and, moreover, contained several Quakers. Thus, it possessed a number of men of military age. However, as Trevelyan noted in his memorandum to the British Ambassador in Rome, Sir Rennell Rodd, these men had become specialists – learning Italian and becoming adept at dangerous Alpine driving. The directive called for the removal of nine ambulance drivers and the adjutant – a loss which would effectively force the unit to cease operations.[73]

After conferring with Dr Brock, Ogston drew up a memorandum outlining the bravery of the drivers and their indispensability to the unit's operations. He also remarked on the esteem with which the hospital was regarded – including its use of female nurses which 'is opening the eyes of the Italian Medical Authorities to the advantages these bring with them'.[74] Ogston sent his memorandum to Sir James Reid – asking it be passed on to Arthur Bigge (formerly Private Secretary to Queen Victoria – a role he continued under George V) 'and, if they thought proper' to the king – and to Walter Long, President of the Local Government Board.[75]

Sir Rodd sided with the unit and wrote to the Foreign Office to stress the political advantages of retaining the unit (the only British presence on the Italian Front at that time). Behind the scenes, however, there was an additional layer of complication. In a further instance of the self-interested interpersonal wrangling Ogston had encountered at the highest levels of voluntary and military medical departments, Lord Augustus Debonnaire John Monson, Red Cross Commissioner for Italy, was

Figure 6.2 Group photo of the Villa Trento staff, preserved in Ogston's WWI journals, under which he wrote the following caption:
Geoffrey Young : Dr Brock : Dyce : Trevelyan : Dr Thompson : Donald Gray
Mr Gilpin : A. Ogston : Mr Harris.
MS-3850-1-8-00165 in the University of Aberdeen Museums and Special Collections, licensed under CC By 4.0.

inclined to use his influence against the unit. Ogston wrote: 'Monson in Italy desires to dominate every activity and benevolence, desires to secure all the credit for his society, and especially seems to dislike this section having its own Committee & its own funds.'[76]

On 24 October, Ogston wrote, 'we held a 'pow-wow' in my room on the burning question of the calling up of the important men of the unit' at which Trevelyan, Dr Brock, Geoffrey Young and others were present.[77] Ogston advised Trevelyan, who had threatened to resign, to avoid committing himself to this course of action until he had 'exhausted every means' of preserving the unit.[78] Trevelyan then left for London to argue the unit's case. In the intervening time, Ogston explored further sections of the front and recorded a visit paid to the villa by the composer and suffragette Ethel Smyth (1858–1944), who was on her way to assist her sister Nina Hollings in her and Countess Gleichen's radiographic work. Smyth composed the women's suffrage anthem 'The March of the Women' (1912) and had, as Ogston remarked, 'been in prison and underwent starvation till dismissed'.[79] Given Helen Ogston's notoriety after she interrupted David Lloyd George at the Albert Hall in 1908, it is possible Smyth would have made the connection between father and daughter. Ogston recalled her as 'a shrivelled old piece of vivacity'.[80]

In November, Trevelyan returned with the news a compromise had been reached: the unit could retain three 'indispensable' staff members and all conscientious objectors, while all others of military age could remain until 'they can be satisfactorily replaced'.[81] 'Meanwhile,' Ogston wrote, 'we have had another sensation':

> A letter from Trevelyan arrives saying Queen Helena called the hospital messy [. . .] passages, yards etc. dirty, food not what it ought to be, & things different from what was to be expected in an English Hospital. He (Trevelyan) commands a great cleaning up, sparing no expense, before the Red Cross Matron, who is to be sent to report on it, arrives.[82]

Head Red Cross Matron Miss Swift arrived on 6 November and began her examination of the unit. Then, on the 9th, Lord Monson arrived at the villa with Hubert Beaumont (sub-commissioner for the British Red Cross in Italy) to report on the matter.

At nine o'clock the visitors came to Ogston's room, along with Trevelyan and Dr Brock. Monson was in favour of separating the unit's hospital and ambulance sections and putting the latter under the charge of 'separate officials'.[83] Trevelyan disagreed. Brock backed him. Beaumont sided with Monson. 'So it went on till 11. Then, as we were coming near a deadlock, Trevelyan asked what I thought. I had avoided saying a word till he asked me.' Ogston remarked that, without the ambulance section, the Villa Trento would change from being a field hospital to a base hospital – which it was not equipped to be. He also argued public opinion would not brook leaving a British ambulance unit without a British medical unit to care for its staff. Although, Ogston did note that, if the ambulance section were required to advance, the accompanying hospital section may have to be reduced and the Villa Trento given over to the Italians. 'This changed the discussion entirely, & finally, though no positive decision was arrived at, there was general agreement that those were the lines to go on.' Ogston also added 'that Queen Elena's reported criticisms were "unjust, & I was prepared anywhere to state this"'. After Monson and Beaumont left, Trevelyan remained in Ogston's room 'pacing like tiger in a cage' until 2 a.m.[84]

Next day, Ogston prepared a memorandum to the effect that the accusations made by Elena 'are baseless, and could not have been [accepted] by anyone acquainted with the facts'.[85] Nearly all the staff signed the document, which was handed to Monson with the statement that it had been originated by Ogston. Ultimately, Monson and Miss Swift decided against recommending the severance of the Villa Trento hospital and its ambulance unit. Throughout these negotiations, one senses Monson's ill-disposition towards the unit would have greater latitude without the outwardly forbidding, august presence of Emeritus Professor Sir Alexander Ogston, KCVO, LLD. Normally sparing of immoderate language, even in his private journals, Ogston described Monson as 'a bounder'. On the same page, he wrote: 'So here ended the first chapter of a nasty job.'[86]

While he esteemed Trevelyan as a scholar, Ogston occasionally found the unit's commander lacked decisiveness. Concerned the hospital would be left behind if the Italians advanced, he pressed Trevelyan to ready the hospital for such an eventuality. Trevelyan 'seemed to assent'. 'But he likes to worry over things a bit & is somewhat too apt to see too many sides to a question, to be a man of rapid action.'[87] Ogston also found Trevelyan's pessimism trying. On 22 November, Ogston noted how, the night previous, Trevelyan had 'got talking about the progress of the world & the reciprocal influences of the war and the higher ideals of humanity, expressing himself sceptical or even pessimistic about the future'. Ogston publicly disagreed with the unit's chief, and the discussion continued later that night in Ogston's room, where, Ogston recalled, 'I freely expressed myself, & we had a long slightly heated argument'.

At breakfast the next day, Trevelyan and Ogston were first to arrive and the former:

> asked if I had forgiven his pessimism of last night, & said he really agreed
> with me in everything. I told him of a friend of mine who enjoyed himself in
> his pessimism and hurt & dismayed his circle by it: that it hurt me to find
> a man of eminence uttering such views & influencing others. That it was
> unpatriotic and almost a crime. He took it right well, & evidently was free
> from 'rancour' about it.[88]

Through the winter, prolonged snow and rain added frostbite, pneumonia, and gangrene to the miseries of the troops. And the personal news reaching Ogston did little to alleviate the gloom. On 23 November, Ogston heard report that the health of Constance – who was honorary secretary of the Aberdeen appeals for sphagnum moss gathering – had 'given way under the strain of her work'. He was also 'very anxious' for Flora, who was 'working quite to the very uttermost limits of her strength'. His youngest son Rannald, meanwhile, was serving at an unspecified location at the Somme or Ancre.[89] Here, Ogston was unusually loquacious on the subject of his children's health. Possibly, entering his eighth decade made Ogston dwell more on the importance of family. In any case, his role in the Great War was not simply that of an eminent surgeon exerting a last opportunity to participate practically in combat: Ogston was also father to a large number of children and must, naturally, have been concerned for their welfare.

During his time at the Villa Trento, Ogston visited numerous Italian base hospitals, concluding, 'it was impossible to form other than a very high opinion of the Italian medical service' which enjoyed, he added, 'a marked absence of the red tape and circumlocution which clings (possibly I ought to say, which used to cling) to our War Office'. One defect he did note was the absence of 'young and competent women nurses'. 'Every Italian, man and woman, to whom I spoke of this subject, told me that the moral standard in Italy forbade the employment of women nurses, and even rendered that of sisters very trying for them.'[90] Prior to Unit One's arrival in Italy, the Italian army forbade the practice of allowing women to nurse in the army. Trevelyan recalls:

> when we first came out, in August 1915, the Italian authorities still had a
> rule against women nurses at the front. At the very moment of our arrival
> this rule was set aside, and we were encouraged to send for women nurses.

The Villa Trento was, then, the test case for female nurses in Italy. Ogston remarked that the Italian patients at the villa responded favourably to this innovation: 'our women nurses were hugely appreciated by them, and theirs was an influence which will have left a lasting impression and may result in helping to modify the practice of nursing in Italian Military Hospitals of the future'.[91]

While conditions at the hospital were demanding, there were reprieves from the pressures of work. During his time in Italy, Ogston visited numerous dressing stations, as well as a field hospital in the Carnic Alps. The English humourist Edward Verrall Lucas (1862–1938) visited the hospital, and his account offers insight into the staff's limited leisure hours: 'In the evening when the work is done the villa turns into a social club. Everyone assembles in a large salon, hung, or rather plastered with large paintings, stuck on the walls in a solid mass [. . .] here are sofas, a piano, English papers and the works of Jack London.'[92] 'In less busy times,' Ogston recorded, 'we were not without amusements.'[93] There were musical evenings, poetry recitals, and tea parties. 'Such evening séances do not, however, last long,' Lucas noted, 'for the villa is a strenuous place and early hours are kept.'[94]

On 19 December, Dr W. E. Thompson – who had previously served as a surgeon at the villa – returned from a leave of absence. This occasioned further anxiety for Ogston, who had been warned of Thompson's occasionally supercilious character. It was agreed that Ogston and Thompson would each be responsible for one surgical and one surgical/ mixed medical ward, with Ogston superintending 'Vittorio' and 'Aosta' and Thompson 'Cadorna' and 'Garibaldi'. Upon initial acquaintance, however, Ogston decided 'there seems no reason why we should not get well along together'.[95] Despite an early reluctance to fault Thompson, Ogston would later find him a disagreeable colleague, writing:

> I don't understand Thompson. His heart is not really in his work [and] he has peevish, petty ways that spoil him. He is at present grumbling that so many British artillery men are patients here, filling up his wards. Yet he has one ward of 16 beds standing empty & he objects to taking any but severe cases into it. We may not have many severe cases to fill it and in the meantime it is not used.[96]

Ogston, however, was no stranger to the corrosive effects of interpersonal wrangling on the happiness and efficiency of colleagues, in peace or war. While he was untroubled by causing offence if this was the unavoidable result of highlighting insupportable deficiencies, Ogston always picked his battles based on their importance to the general good. For all his other faults, much of the dignity in Ogston's character stems from the incontrovertible truth he was consistently motivated by the gravity of the task in hand – and never by a desire for recognition of his own significance.

A special effort was made for Christmas 1916, at which dinner – accompanied by the villa's own wine – was served to eighty guests. Afterwards, the diners went to the lounge to watch a three-act play devised by Geoffrey Young. The first act showed the comic attempts of VADs in London attempting to gain admission to the unit by bribery; the second lampooned the eccentricities of the unit's members, including Trevelyan (though not, it would appear, Ogston); the third act jumped ahead sixty years to an imagined scene where four men of the unit, now in their dotage, were cared for by an aged nurse. At this point Father Time (Young) entered carrying a scythe, which he threw down, thereby announcing the war was, at last, over. This was celebrated with a comic song to the tune of 'The Duke of Plaza Toro' from *The Gondoliers*. As a further mitigation to spending Christmas in a mist-bound warzone far from home, the fruits of Ogston's ambulance appeal appeared. On 26 December, he jubilantly noted the arrival of '4 ambulance cars, results of my appeal to Aberdeen arrived today, 2 from Hector Cameron (Glasgow Red Cross), one from Mrs. Smith, Aberdeenshire Red Cross and one from Lady Sempill'.[97] A few days later, on New Year's Evening, there was a dress ball. Some of the staff were keen to see Ogston wear a kilt. 'I was deaf to the seduction,' Ogston wrote, 'being too old and grave for such a piece of folly.'[98]

Such rare moments of relaxation were vital to the functionality of the staff. In addition to long hours and demanding – often dangerous – work, the unit staff faced other burdens. In midwinter the landscape was 'swept eternally [by] misty veils of rain'.[99] The only fuel – damp logs – provided little alleviation against the persistent chill. In January, the mists lifted to be replaced by 'snow, frost, and wind [. . .] of the most biting kind'.[100] Water pipes froze and sanitary arrangements became challenging. 'One had to keep a tight rein,' Ogston remarked, 'to avoid thinking and saying nasty things.'[101]

On 29 December, Dr Brock left for a month's holiday, placing Ogston in charge of the hospital in his absence. Ogston had to remain in the hospital at all times, seeing every new casualty, greeting visitors, and caring for the sick and wounded. In addition to these strains, Brock's absence was a social deprivation for Ogston:

> I miss Brock greatly. Trevelyan and I never have been and are never likely to be intimate, he is such a moody man, & the others, with exception of Geoffrey Young in Goritzia, are so much younger than I that great intimacy is improbable.[102]

During Brock's absence, Trevelyan visited Ogston in his room, and the pair discoursed at length on prehistory, warfare, socialism, and education. This exchange vivified the differences between these two distinguished individuals:

> It was pretty one sided, for Trevelyan talks well, and thinks better, while the very converse is the case with me, so that my part was throwing in suggestive remarks, & somewhat guiding the shift of the conversation. When speaking with him, I have always the sensation as if I were handling a loaded, cocked revolver. But he is an uncommon man, & most interesting.[103]

Lack of a cordial relationship with the unit's commander was one of many burdens to be shouldered by Ogston, who was responsible for the disinfecting and quarantining of the hospital's diphtheria patients, dietary regulations, and – in addition to his own surgical cases – was consulted 'on every difficult medical and surgical case' in the hospital.[104]

Worse, two of the hospital's doctors were called away, leaving only Ogston and Thompson (who spoke no Italian) to manage 100 in-patients and 'out-patients, perhaps 60 a day, and all the casual cases that turn up all the time'. Furthermore, owing to illness, the hospital was short of nurses. Trevelyan dealt with this news by brooding, in silence, in Ogston's room for quarter of an hour, only to abruptly depart then declare later, at dinner, that Brock must be recalled from holiday. Ogston was strongly opposed to this and, eventually, the staff agreed to his plan of closing the outpatient department, barring new admittances, and rearranging the wards. 'Dear Dr. Brock ought to bless me,' Ogston wrote.[105]

On 29 January, waiting in the freezing station square to greet Brock, Ogston's thoughts turned to his son Rannald 'in command of his own battery of siege guns in France'. Brock was not on that train, however, nor the next. On returning to the villa, it was discovered he had telegrammed to say he was ill with influenza and therefore unable to travel. Ogston lamented the meticulous handover reports he had prepared were now useless. His next remark, however, further elaborated on interpersonal strains only hinted at in Ogston's *Reminiscences of Three Campaigns*. 'I am in for at least a week more of the Directorship of the hospital, & the association of that icicle of a man, Trevelyan. He is an icicle: clear, straight, & clever, but so cold. An icicle one can not but respect, but without the feeling amounting to liking. Well! Well!'[106]

To Ogston's relief, Brock returned to duty on 7 February. It was suggested Ogston have a month's holiday. He decided this would be impossible, however, as the unit had no second surgeon and cases were 'hourly coming flooding in'.[107] Despite Brock's return, the unit was far from harmonious. There was some uncertainty as to the hospital's future. Would its staff be relocated? Or would it remain in situ, perhaps in anticipation of intense fighting on the Carso? Trevelyan, meanwhile, was becoming discouraged and stated his disinclination to remain another winter. His pessimism shaped wider perception of the hospital's and ambulance unit's future, 'causing the members look on it as best that the unit should collapse in Autumn'.[108]

With the arrival of spring, preparations for the coming offensive began, spreading a welcome expectancy among the unit. In April the British Artillery arrived and established ten batteries on the Carso. On the eve of his seventy-third birthday, Ogston appreciatively recorded receiving the gift of a pipe and 'my old smoking mixture' from Flora.[109] Finally, in May, the weather improved, incipient grapes appeared on the vines, and military 'activity behind the front became positively delirious'.[110] The warmer weather also brought with it scorpions, mosquitos, and tarantulas that 'swept like shadows across the walls of our rooms'.[111] As it did in the Boer War, Ogston's health declined. Suffering from dysentery, he could only continue to function by dosing himself with opium (chlorodyne), and in this 'half-narcotised condition, and when a hot muggy sirocco was blowing, which made one hot and sticky' Ogston had to draw up a 'complete and connected scheme of mobilisation in Sections' for the hospital's possible advance.[112] Elsewhere, preparations were afoot. The Italians built a new road down to their bridgehead on the Isonzo River which was to be the base for the assault on Monte Kuk. The unit maintained an ambulance outpost here which operated under the command of Philip Noel-Baker. From 15 to 30 May 1917, the ambulances stationed at Plava carried over 8,000 casualties.[113]

The heightened military activity occasioned a corresponding rise in the hospital's workload. On 15 May, Ogston noted:

> our cars commenced to bring in wounded from Monte Santo, & ever since an increasing stream has been arriving, so that from being absolutely empty [the hospital] has become half full, shell and bullet wounds almost exclusively.
>
> At 3 this morning, a large batch came, at 5 another, at 9.30 a third, and since morning the flow has been steady. Unfortunately I have got a touch of lumbago, at any rate a crick in my back, but it did not hinder my giving full attention to twenty-three freshly wounded men, and getting them dressed and comfortable, though it tired me somewhat.[114]

On the 26th Ogston recorded performing: 'My 60th operation, and first death, from shell wound of the hip joint, with pyaemic, secondary haemorrhage, necessitating disarticulation at the hip joint.'[115] June and

July saw extreme thunderstorms and high humidity. Conditions began to tell: 'there was much sickness among the members of the unit; some collapsed from diphtheria, others from typhoid fever; there were even slight attacks of sunstroke, and nearly all were weakened by dysenteric or abdominal complaints'.[116] Personally, Ogston conceded:

> I am not now a very strong man in some ways. I am just able to carry on, but feel that little would upset me, & in many ways I can see the possibility of my failing to stand the work, soon must admit that I might break down & have to go home or even I might die here.

In Ogston's assessment, Dr Thompson – though the younger man – was less physically robust of the pair, and noted 'if either Thompson or I were to fail, it would be impossible, or nearly so I fear, to get anyone to take our place'.[117]

Having captured Monte Kuk, the Italians continued to assault Monte Santo, San Gabriele, and the Austrian positions on the Carso. In August, Geoffrey Young was wounded in the left thigh by a shell-burst; after gangrene set in, Young's leg was amputated. Another driver, Lionel Sessions, also lost a leg. The noise of planes passing over the villa became so loud and continuous it interfered with stethoscopic examinations. Meanwhile, military traffic incessantly rumbled to and from the front in the August heat. By this point, Ogston had served for over a year with only three days off due to sickness. In September 1917, he left for a six-week furlough to restore his energies for the remainder of the conflict.

On 18 September in Milan, Ogston received a telegram informing him his brother, Frank, had died suddenly in New Zealand, where he was working as a sanitation officer.

> Poor Frank, when last I received a letter from him, he had set his heart on retiring from work & coming home to Scotland to spend the remainder of his days among us. He had been writing that he had suffered from asthmatic attacks, & that letter mentioned that he had been feeling his heart troubling him. No doubt it was his heart which caused his death. These are no times for troubling others with one's own private griefs, and I mentioned the contents of the telegram to no one, and I have put on no sign of mourning and have no wish to do so. There is immense mourning enough going on amongst us, & the outward "trappings" are better dispensed with.[118]

Arguing personal grieving was unseemly at a time of global conflict may have been an attempt by Ogston to rationalise – and thereby alleviate – this bereavement. In any case, while facing this latest personal loss,

Ogston again exhibited the stoicism which was such a marked element of his character.

In London, Ogston called at the Pall Mall office of the First British Ambulance Unit to Italy and visited his old hospital at Southall, which – at Ogston's suggestion sent via Sir James Reid – had since been visited by the king and queen. The long journey to England and various visits he paid to his family in Britain on his return exhausted Ogston. On reaching Aberdeen he took a week of quiet rest – declining even to travel to shoot at Glendavan with Rannald.[119] Somewhat revived, Ogston was glad to catch up on family news. His daughter Constance was leaving to serve with the National Service Ministry, while Flora had been busy with the City War Work Association. Walter managed to arrange some time off and spent a weekend with Flora and Ogston at Glendavan. Ogston enthused, 'the glory of being once more out on the moors, and among the woods & the plantations beside the water of the loch! Words cannot tell it'.[120]

On 24 October, the Italian advance was dramatically reversed when German forces breached the Italian lines at Caporetto, precipitating the Italian army's chaotic retreat to the River Piave. Ogston, who had visited the Italian lines north from Gorizia to Caporetto, remarked he 'was invariably struck with the fact that, while the Isonzo front was enormously fortified, there seemed to be comparatively little or no provision for the defence of Caporetto'.[121] The Villa Trento staff were caught up in this retreat and the hospital abandoned. Ogston received news the hospital personnel had safely fallen back to Castelbelforte. Eager to rejoin them, Ogston was detained in Aberdeen, as the Director of the Northern Bank had asked him to chair the annual shareholders' meeting. Having done so on 2 November, Ogston took the sleeper to London.

Ogston called again at the headquarters of the British Committee in Aid of Italian Wounded. There he and met its Chairman, E. H. Gilpin, who asked Ogston to reassure Trevelyan that the Committee was willing and able to offer any necessary financial assistance.[122] Pressing on to Paris, Ogston happened to encounter members of the unit as they poured into the lounge of the St James's Hotel alongside a crowd of refugees. The unit members possessed only the clothes they wore – excepting a kitten which one of the female staff had carried all the way in a basket. Ogston journeyed on to look for the others, travelling via Modane, Turin, and Milan to Castelbelforte, where he caught up with the remnants of the Villa Trento staff. His colleagues, 'starving for something to read in the long dark winter evenings', gratefully received the box of books Ogston had taken with him.[123]

Some of the unit in Castelbelforte were sick. Finding a lack of medical supplies and personnel, Ogston, now seventy-three years of age, went to Mantua to source disinfectants and invalid food. The unit's staff were billeted in a house whose owner had packed all his possessions in anticipation of a further Austrian advance. Here Ogston was disappointed to learn of Trevelyan's decision that the Villa Trento unit was to be disbanded. 'I could hardly credit this,' Ogston wrote, 'when so many of our own members and other British were liable to be taken seriously ill.' Furthermore, Ogston was disappointed to hear that his offer to 'continue a small well-equipped hospital' was refused.[124] These misfortunes were rendered more acute by Trevelyan's disreputable conduct following the Caporetto retreat. It was not obviously within Trevelyan's purview to conclude the hospital part of the unit and disband its staff, which he did without consulting Dr Brock or the unit's London Committee. More ignominious still was the news Trevelyan had decided to terminate Brock's service and instead retain Dr Thompson. Not only had this decision been reached without consulting Brock, with the unit's hospital discontinued, Thompson would only have to attend a small number of healthy men comprising the unit's ambulance division. As Brock expressed himself in a letter to Ogston: 'It seems to me not quite the thing for a young man to do [. . .] for he ought to be attending to the wounded instead of wasting his talents on a motor yard.'[125] Ogston offered his services to the British Red Cross Commissioner, but this was also declined. He remained in Castelbelforte until arrangements were made for the wounded he had been supervising. Then – having been awarded the Italian Long Service medal and the Italian Silver Star – Ogston travelled back to Scotland on 17 November 1917.

In Aberdeen, having received letters from Brock, Gilpin, and Trevelyan confirming the discontinuance of the unit's hospital section, Ogston sent a letter to Bevan Baker, Honorary Secretary to the First British Ambulance Unit for Italy, in which he formally severed ties with the organisation. Dated 26 November, the letter concluded: 'It would not be easy for me to find words to adequately express the reluctance with which I sever my connection with your Unit and the many friends I have made within it, but the letters I mention leave no hope that things may be otherwise ordered.'[126] Having spent so long in the warzone, the domestic quietude of 252 Union Street must have magnified Ogston's sense of dislocation from the current of great events then unfolding. The 1917–18 volume of *Aberdeen University Review* records Ogston was elected to the Joint Committee representing the Department of Voluntary Organisations and to the Scottish branch of the Red Cross.[127]

Herbert Grierson claimed he suggested that, upon his return from Italy, Ogston work up the journals from his military campaigns into an account of those adventures. Grierson recalled assisting Ogston with this project in the summer of 1918 during editing sessions held in the summerhouse at Glendavan.[128] It is intriguing to imagine Grierson and Ogston sitting together amid the birdsong and pine trees of a Deeside summer while Allied forces, bolstered by continual arrivals of American troops, successfully turned the tide of the German Spring Offensive. Did surveying a lifetime of useful action provide Ogston some measure of satisfaction which offset his disappointment of experiencing the war's final phases as a remote spectator?

By the time World War I drew to a close the following November, approximately 9.2 million combatants had been killed. Far deadlier than battlefield casualties, however, was the 1918–19 flu pandemic – in part spread by troop movements – which killed between 20 million and 100 million people, depending on estimates.[129] Many of those who died did so not from the influenza virus itself, but from secondary infections, including bacterial pneumonia caused by *Streptococcus* and *Staphylococcus aureus*, the bacterium Ogston had first observed in his garden laboratory in Aberdeen nearly forty years before.[130]

A few months after Ogston's death, aged eighty-four, on 1 February 1929, the American novelist Ernest Hemingway published *A Farewell to Arms*, one of the most important works of American twentieth-century fiction. Ogston would have been interested in Hemingway's novel, as it is set in north-eastern Italy during World War I – in the very locations and timeframe Ogston personally experienced. Indeed, to anyone familiar with Hemingway's novel, Ogston's Italian reminiscences immediately recall *A Farewell to Arms* in a number of ways. A British hospital housed in a villa, an ambulance unit in Gorizia, feverish roadbuilding, preparations for an attack at Plava, epidemic cholera, roads hidden from the enemy by screens, the hospital's involvement in the chaotic retreat from Caporetto – all these details from the Italian section of the *Reminiscences* overlap with *A Farewell to Arms*.

In 1918, approximately seven months after Ogston returned to Scotland, the young Hemingway travelled to north Italy as a volunteer ambulance driver with the American Red Cross. Hemingway served at the front for only a few weeks, as he was injured in a mortar blast on the Piave River, but he would return to the war repeatedly in his fiction. *A Farewell to Arms* deals with a young American ambulance driver, Frederic Henry, falling in love with a nurse named Catherine Barkley, who works at a hospital on the Italian Front. Although Hemingway's experiences informed the novel's war scenes in general, *A Farewell to Arms* is set on the Isonzo Front, in Gorizia and the vicinity of the Villa Trento hospital – a part of Italy Hemingway had not visited at the time he was composing the novel. The events of *A Farewell to Arms*, furthermore, occur during, and in the build up to, the retreat from Caporetto, which Hemingway arrived in Italy too late to experience.

In order to write about these events with convincing specificity, Hemingway supplemented his personal experience of the Italian Front with specific details gleaned from memoirs and history books. Chief among the sources he consulted was G. M. Trevelyan's 1919 memoir, *Scenes from Italy's War*.[131] Trevelyan's work contains details concerning transportation, climate, disease, and geographical particulars that Hemingway mined for his highly accurate depiction of the war on the Gorizian Front from 1915–17. As the Villa Trento was the only British Red Cross hospital on the Gorizian Front during the war, the hospital formed the basis of Hemingway's depiction of the 'British hospital' at which Catherine and Frederic meet in the novel.[132] Trevelyan's description of the Villa Trento as an 'old eighteenth-century Schloss' (the German word for manor-house) may have influenced Hemingway's decision to give his British hospital a German origin.[133] Hemingway alludes twice to the British hospital's original German owner: 'The British hospital was a big villa built by Germans before the war,' and again: 'This had been the villa of a very wealthy German.'[134]

Of course, this does not mean the British Red Cross hospital in *A Farewell to Arms* is entirely based on the Villa Trento. In fact, in a deviation from historical accuracy, Hemingway places his hospital in Gorizia. There had, at one point, been plans to relocate the entire Villa Trento staff to Gorizia, but these were never realised, partly, as Ogston remarked, because of a 'reluctance to expose our women nurses to the greater risks of hostile fire there'.[135] Michael Reynolds contends Hemingway 'took liberty with the facts in order to move Catherine into close proximity with Frederic in Gorizia'.[136] As the only British Red Cross hospital in the vicinity, however, it is highly likely Hemingway had the Villa Trento in mind when describing the 'British hospital' in *A Farewell to Arms*.

In addition to the fact it was the only British Red Cross hospital in the area, there is a further detail which links the Villa Trento and the 'British hospital' of *A Farewell to Arms*. One evening, the year following his move into Gorizia, Frederic visits Catherine Barkley. Citing another example of Hemingway's historical accuracy, Reynolds notes: 'one reason that Catherine Barkley is a British VAD is that in 1917 there were no American Red Cross nurses at the front in Italy'.[137] However,

Hemingway's knowledge of historical facts goes a step further in regard to Catherine's status as a female medical worker on the Italian Front. On Frederic's second visit to Catherine at the 'British hospital' she tells him: 'The Italians didn't want women so near the front. So we're all on very special behavior. We don't go out.'[138] The tension generated by the newly permitted presence of women nurses on the front places a degree of constraint on Frederic and Catherine's initial courtship and further links the Villa Trento to the hospital at which they meet in *A Farewell to Arms*.

By the time of his meeting with Catherine, Frederic has been serving with the Italian army for two years. While his wounding and escape from execution at the Tagliamento River may prove traumatic, Frederic's first two years, almost entirely glossed over in a compressed narrative lasting just a few pages, should also be taken into consideration. Descriptions of first-hand witnesses, like Ogston and Trevelyan, provide a fuller picture of the conditions Frederic would likely have encountered at the front, but which Hemingway almost completely omits to describe. Frederic claims that his duties at the front 'seemed no more dangerous to me myself than war in the movies'.[139] Yet, while he implies his duties are not taxing, being an ambulance driver on the Isonzo Front would have been an often demanding and even perilous experience.

On one occasion, a shell fell through the roof of the unit's main Gorizian ambulance station, seriously wounding the only person inside.[140] As Ogston notes, many ambulance convoys travelled at night to avoid 'shelling and bombing by aircraft'.[141] As no lights were allowed on the road (in case of alerting the enemy to their position), drivers had to manoeuvre in the near-darkness, leading to 'big smashes occurring'.[142] In his memoir, Geoffrey Young recalled driving in these conditions: 'At last a darkness was close round my face that felt like a black velvet mask without eye holes. I could see nothing at all [. . .] I shut my eyelids on the blackness, and whether I had them shut or open it made no difference at all!'[143] 'Even by day,' Ogston remarked, 'the ambulance driver's work was sometimes no light one; they had to bring their freights of wounded, during the winter months, through veils of fine driving snow which penetrated every cranny and article of clothing.'[144]

Following Frederic's recovery in a Milanese hospital, he returns to his ambulance unit in Gorizia and is sent up to the Bainsizza Plateau. It is here that Frederic finds out the Austrians have broken through at Caporetto in the north. After evacuating field hospitals on the plateau and a clearing station at Plava, Frederic returns to Gorizia, where he packs the ambulances under his control with medical equipment and joins the retreating troops. Like Trevelyan, Frederic turns off to the north in an attempt to circumvent the blocked roads during the retreat,[145] while both narratives record *carabinieri* being present on the Tagliamento bridge to assort the retreating men.[146]

A Farewell to Arms has become a canonical novel of World War I – a conflict known for the impersonal slaughter of combatants via the means of poison gas, heavy artillery, and machine guns. Warfare had changed vastly in the thirty years since Ogston marched in square formation with the First Bearer Company at the Battle of Hasheen. While he was stationed on an Alpine front – away from the mud and more level topography of northern France – the wounds Ogston treated were largely inflicted from an unseen enemy: 'Most of the wounds we had to treat were from shell and grenades [. . .] the minority were from shrapnel and rifle bullets; and there was not a single case of bayonet wound among those which came under my personal observation.'[147] 'I was not sorry to have the opportunity of seeing what modern war was like,' Ogston wrote, noting, '[i]t was very different from my former experiences.'[148]

Figure 7.1 Portrait of Sir Alexander Ogston, oil on canvas by George Fiddes Watt ARSA, 1910. ABDUA:30123. Image courtesy of University of Aberdeen.

7

1919–1929

As we turned one corner and came into the view of Loch Davan and the
hill Morven beyond, something in the air, perhaps in the look of the heather
and the trees, brought it vividly before my mind that summer was over.
  'How quickly time flies.'
  'And life with it,' was Ogston's slightly melancholy rejoinder.
Herbert Grierson[1]

O N 8 FEBRUARY 1919, Ogston's son Alexander Lockhart Ogston – a
captain in the Royal Marine Artillery – died of pneumonia, aged
thirty-one. Ogston had now lost sons in the Boer War and Great War.
Perhaps this added a sombre note to any satisfaction he may have derived
from the publication of *Reminiscences of Three Campaigns* in 1919.
Dealing with Ogston's time in the Soudan, Boer War, and World War I,
Reminiscences was described in the *BMJ* as being 'much more than a
record of observations and opinions on medical matters: the indomitable
energy of the Emeritus Professor of Surgery in the University of Aberdeen
[. . .] led him in the interval of his strictly medical duties to contrive to see
a good deal of the actual military operations in all the three campaigns'.[2]
And it is an undoubted strength of the book that Ogston's willingness to
explore the warzone – in reconnoitring the British lines of attack before
Lord Roberts's arrival in Cape Town, for example – lends his account
a more expansive overview of the military medical scene than another
surgeon working in the field may have produced. In recognition of his
Italian service, in January 1920 Ogston received the Cross of Cavaliere
dell'Ordine della Corona d'Italia (the Order of the Crown of Italy).
   At the BMA's 1922 Annual Representative Meeting, the Chairman
of the Representative Body brought forward a recommendation that
Ogston – who had been BMA President through 1914–16 – be elected

Vice-President. Ogston was duly accorded this role, one which would have been 'brought forward earlier' if not for the war.[3] Age, however, began to catch up with even one so vigorous and determined as Ogston. He became less active physically as a result of an arthritic hip bone, gave up spending time at Glendavan, and remained in Aberdeen, relying more on the society of his children. Walter noted with regret that, being engrossed in his own career and family, he 'in a sense neglected' Ogston, who displayed 'a certain pathetic yearning [. . .] towards me, particularly during his last few years'. Walter did manage 'to visit him several times a year for week-ends, but they were very quiet ones', involving 'a cycle ride round the neighbourhood of Aberdeen or, when cycling had to be given up, a drive'. At 252 Union Street, the visits were spent reading, playing patience, or conversing, '[b]ut talking became less as time went on'. Noting his father's disdain for speaking unless there was something to say, Walter remembered, nevertheless, that Ogston 'loved to have Flora and me with him'.[4]

On 27 June 1923, Ogston was too ill to attend a dinner given at the Palace Hotel in Aberdeen to honour Henry Gray (1870–1938). Gray had been Ogston's Resident Surgeon at ARI, and Ogston greatly esteemed his abilities, remarking: 'I do not think such work as he has given could be surpassed by any one.'[5] While Ogston had been an early adherent of antiseptic surgery – which used carbolic acid to destroy bacteria in operative wounds – Ernst von Bergmann pioneered a system of aseptic surgery which used steam to sterilise surgical instruments and dressings. As a postgraduate student in Germany, Gray became familiar with these aseptic techniques, and, upon becoming a consultant surgeon at ARI in 1904, was able to bring aseptic practice to Aberdeen. Gray was also instrumental in introducing the use of local anaesthesia to Britain. During the Great War, meanwhile, Gray realised gas gangrene was best treated as soon as possible by early excision of contaminated tissue, thereby greatly reducing the impact of septic wound infections.

In 1923, on his way to Montreal to take up the post of Surgeon-in-Chief at the Royal Victoria Hospital, Gray told the Aberdeen gathering:

> He regretted very much that illness had prevented Sir Alexander from being there, for he should have liked to have acknowledged to him personally, before them all, the great debt he owed him for the inspiration, stimulus, encouragement, and many kindnesses he had received at his hands.[6]

By November, Ogston was well enough to deliver as his presidential address to the Aberdeen Medico-Chirurgical Society a recent history of anaesthetic methods. Ogston remarked that, in Aberdeen, the usual practice was to administer a chloroform anaesthetic. 'About 1917,' however, 'open ether [anaesthetic] came into more common use, having been brought to the notice of anaesthetists by Sir Henry Gray after a visit to America.'[7]

In his last years, Ogston suffered a stroke, after which he was cared for by his daughter Flora and a nurse. His grandson Alexander G. Ogston, notes that by 1926, Ogston was bedridden.[8] Both his son Walter and Herbert Grierson remarked this was a particularly tragic end for such a dynamic man. Walter, in fact, was certain his father had wished to meet his end in the Italian warzone. Instead: 'The pathos of his life was his end, the long, gradual, drawn-out dissipation of his physical and mental powers, until even his speech grew confused. It was the very reverse of what he had longed for.'[9]

In his last days, Ogston refused to be seen by anybody bar his own children, although he made an exception for his newly born great-grandson, William. On this visit to introduce her son to his eminent great-grand-father, William's mother, Mary Grierson, accidentally saw into Ogston's room while walking in the garden at 252 Union Street. She remembered the incident: 'I saw Grandpa for the last time. He was lying propped upright in his bed, very soldierly and fine, his old features composed in a long patience, full of undaunted courage and resolution. He had fought the fight, he had run the race, and now was waiting in the dignity of his own soul for the last summons.'[10] Ogston died on 1 February 1929, aged eighty-four, and was buried in St Clement's churchyard, Aberdeen, in the same grave as his first wife, Molly.

The *BMJ* obituary contained an appreciation from Ogston's successor in the Chair of Surgery at Aberdeen University, Sir John Marnoch, who noted to:

> those of us who had the privilege of being his pupils when he was at the zenith of his career as a surgeon and teacher he made a profound impression. He was the most outstanding personality in the University.
>
> At the time I am thinking of, in the late 'eighties, his researches into the cause of suppuration had earned for him a world-wide reputation, and it was therefore with great reverence, not unmingled with awe, an awe that lingered with us even in later years, that we approached him as the great master of the science and art which we had chosen to learn. His great powers of deft and graphic description made his lectures delightful to listen to. [. . .] His course of Systematic Surgery was much in advance of anything found in the standard text-books of the day.[11]

As the outcry generated by Ogston's announcement of his decision to retire in 1892 indicates, he was a revered and respected teacher. Not least among the many accomplishments of his career was his influence as an educator at Aberdeen University and as a mentor to those working as his assistants at ARI. T. Wardrop Griffith, Professor Emeritus of Medicine at Leeds University, remarked Ogston's death would 'cause a great many of his former students [. . .] to recall to mind an old teacher with feelings of admiration, respect, and affection'.[12] In addition to Sir Henry Gray, Sir John Marnoch, and Professor Griffith, Ogston taught – or had as an assistant – Sir James Mackenzie Davidson (1856–1919), one of the first to produce X-rays in Britain; Arthur R. Cushny (1866–1926), Professor of Pharmacology at University College, London, and later Professor of Materia Medica and Pharmacology at Edinburgh University; William Bulloch (1868–1941), Professor of Bacteriology at London University; and Sir Ashley W. Mackintosh (1868–1937), Professor of Medicine at Aberdeen University and Honorary Physician to the King's Household in Scotland. Meanwhile, among the reminiscences of Ogston provided by family and friends which Walter edited together for *Alexander Ogston K.C.V.O.* (1943) are numerous affectionate tributes from former pupils who entered the medical profession.

Walter acknowledged his father's 'closing days were sad' and that '[p]ossibly he lived too long'. Equally, he considered his father had lived 'a very long life, a very full life' and, Walter was 'convinced, a very happy life'.[13] Constance, meanwhile, remarked her father:

> was a great man and a wonderful father, and his mind and character were of such a stimulating quality that when his children think of him there is suggested to their minds something lofty and elevated, resembling [. . .] the wild scenery where he built his country home – with rolling heather moors, unfettered streams plunging over rocky beds, and the indomitable hills.[14]

Situated among pasture lands, heather moors, broad-backed hills and distant mountains, the environs of Glendavan – the country home Ogston built in Aberdeenshire's Dee Valley – form an apt analogue to represent the man. The wide views the Dee Valley affords, of river, field, moor, hill and mountain, suggest the scope and variety of Ogston's intellectual life, the gravity and solidity of his character, and the permanence of his achievements.

Ogston's career, like that landscape, climbed to great heights. Equally, many of the charms of Deeside unfold on a more modest scale – in agricultural, plant, animal, and insect life. Likewise, many of Ogston's most pleasurable pursuits were experienced at a local, domestic, and unpretentious level. He had a large family, several hobbies – reading, learning languages, photography, archaeology, sketching, cycling, and croquet – and loved smoking. Moreover, the association between Ogston and the Dee Valley is more than analogous: Ogston's life and that landscape literally overlapped in the shooting, fishing, walking, and family reunions the Ogstons enjoyed during numerous summers at Glendavan, in the research Ogston carried out into local prehistoric remains, and in his service as a royal surgeon at Balmoral. Grierson records: 'During the last rather sad years of his life he told me once that the greatest pleasure he had at Glendavan was sitting in the little summer house, and listening to Mary and Flora talking together.'[15]

Quick to grasp the significance of Lister's use of carbolic acid, Ogston brought antiseptic surgery to Aberdeen. Among the remarkable advances in antiseptic practice created in elective surgery, Ogston added his operations for *genu valgum* and club foot (in both adults and infants). He also contributed to the understanding of joint anatomy by discovering that articular cartilage is, rather than a form of extraneous padding, an integral part of bone structure. Most notably, Ogston discovered *Staphylococcus* and conclusively linked this to the acute suppuration and infection that blighted virtually every post-operative wound during the days before antiseptic surgery. Moreover, he successfully defended these findings in the face of criticism from Joseph Lister and, taking them a stage further, concluded blood poisoning – rather than resulting from 'bad' or infected blood – was, in fact, the result of localised infection spread throughout the bloodstream by micrococci.

Surgery and bacteriology alone were not enough to confine Ogston's energies, and his investigations into military medicine – which brought him to Germany, Russia, the Soudan, South Africa, Serbia, and Italy – earned from Surgeon Vice-Admiral Sir James Porter (1851–1935) the high praise that Ogston was 'the first pioneer in dealing with the medical conditions prevailing in the Army and Navy in modern times'.[16] Ogston was instrumental in bringing about the foundation of the RAMC, and his subsequent personal experience in the Boer War amply confirmed that the criticisms he had made of the Army Medical Department were in no sense overstated. His appointment as Queen Victoria's Surgeon in Scotland, presidency of the BMA, awards of LLD degrees from Glasgow and Aberdeen universities, and knighthood were honours appropriately acknowledging and extending a career of far-ranging importance throughout the fields of surgery, bacteriology, education, and military medicine.

Ogston had ten children, and he was survived by many distinguished descendants. One grandson, Alexander George Ogston (1911–96), rounded off an impressive career as a biochemist by serving as President of Trinity College, Oxford, from 1970 to 1978. The grandchildren produced by the union of his daughter Molly and Grierson were also notable. Flora Grierson (1899–1966) published a history of Edinburgh, *Haunting Edinburgh* (1929), and co-founded the Samson Press, which published works by Willa and Edwin Muir. Janet Teissier du Cros (1905–90), meanwhile, wrote *Divided Loyalties: A Scotswoman in Occupied France* (1962) about her time in France during World War II, and a memoir concerning her early life, *Cross Currents: A Childhood in Scotland* (1992).

In 1980, on the centennial of his discovery of the germ of acute suppuration, Ogston was honoured by an international conference held in his name at Aberdeen University on the subject of *Staphylococcus* infections. In 2002, meanwhile, The Ogston Society was founded at Aberdeen University. The society encourages aspiring surgeons by providing practical and theoretical training for undergraduates at Aberdeen. And portraits of Ogston still hang in the Aberdeen Medico-Chirurgical Society's lecture room and in Elphinstone Hall, where graduations are held at Aberdeen University. In 2022, Aberdeen City Council installed a commemorative plaque at the site of Ogston's former home at 252 Union Street. The inscription reads:

Sir
Alexander Ogston
(1844–1929) Professor of Surgery,
surgical innovator and
distinguished military surgeon
lived in a house on this site
from 1870 to 1929.
In 1881 Ogston discovered the Staphylococcus organism in
a laboratory he constructed in his garden.[17]

Ogston's death brought to a close a long and remarkably distinguished life. He had brought antiseptic surgery to Aberdeen, discovered *Staphylococcus*, co-founded the RAMC, and become a trusted surgeon to the royal family. His travels took him to the deserts of Soudan, the storm-ravaged wastes of South Africa, the side of Tsar Nicholas II in Russia, and to the artillery-illuminated night sky of northern Italy. He had seen war change from the days of cavalry and marching in square formation to the fully evolved modern war of the trench and machinegun. More than being an observer, Ogston offered his considerable skills to easing the suffering of his comrades in the field, and by tirelessly campaigning for improvements in the infrastructure of British military medicine. His achievements had incalculable benefits and stand as a comprehensive rejoinder to the disparagement offered by the editor of the *British Medical Journal*. Ogston's career does not, however, shine out only in the context of his native city. His achievements place him, alongside Joseph Lister, Robert Koch, and James McGrigor, as one of the foremost contributors to surgical, bacteriological, and military medicine of his age.

Notes

CHAPTER 1

1. Alexander Ogston, 'How Antiseptic Surgery Came to Aberdeen', in *Sir Alexander Ogston K.C.V.O.*, ed. by Walter Ogston (Aberdeen: Aberdeen University Press, 1943), 93–7, 95.
2. Alexander Ogston, 'President's Address Delivered at The Eighty-Second Annual Meeting of the British Medical Association', *The British Medical Journal*, 2(2796) (1914), 221–8, 222. JSTOR.
3. Ibid. 226.
4. Cambridge University was founded in 1209 and teaching at Oxford University dates to 1096.
5. Alexander Ogston, 'President's Address Delivered at The Eighty-Second Annual Meeting of the British Medical Association', 221.
6. Ibid. 226.
7. Some ambiguity exists as to whether the sobriquet 'Soapy Ogston' was applied to one of the manufacturing 'Alexander Ogstons' in particular, or if it was passed from father to son. It is possible the name may also have passed into local vernacular as a general reflection of the family's wealth.
8. John A. Henderson, *History of the Parish of Banchory-Devenick* (Aberdeen: D. Wyllie, 1890), 95.
9. Ibid. 96.
10. 'Norwood Hall Hotel', British Listed Buildings, https://britishlistedbuild-ings.co.uk/200406334-norwood-hall-hotel-garthdee-road-aberdeen-peter-culter#.YNhckehKg6Y, accessed 17 July 2022.
11. 'Kildrummy Castle', Historic Environment Scotland, http://portal.historic environment.scot/designation/GDL00237, accessed 17 July 2022.
12. Alexander Ogston, *The Prehistoric Antiquities of the Howe of Cromar* (Aberdeen: Third Spalding Club, 1931), v.
13. A. H. Millar, and Brenda M. White, 'Ogston, Francis (bap. 1803, d. 1887), expert in forensic medicine', *Oxford Dictionary of National Biography*, Oxford Dictionary of National Biography (online).

14. In 1869 Malloch established a practice in Hamilton, Ontario, where he also worked at the City Hospital. See: James Kirk Houston, 'An appreciation of A.E. Malloch, MB, MD (1844–1919): a forgotten surgical pioneer', *CMAJ: Canadian Medical Association Journal* 160(6) (1999), 849–53.

15. Carolyn Pennington, *The Modernisation of Medical Teaching at Aberdeen in the Nineteenth Century* (Aberdeen: Aberdeen University Press, 1994), 7.

16. Alexander Ogston, 'Personal Record', in *Sir Alexander Ogston K.C.V.O.*, 53–62, 53–4.

17. These include an annotated timeline of Ogston's education and professional career, accounts of his student days, and sections on antiseptic surgery in Aberdeen, the discovery of *Staphylococcus*, and meetings with Queen Victoria. Ogston's son-in-law Herbert Grierson notes that, after Ogston's return from World War I, he encouraged his father-in-law to publish accounts of his military adventures. These appeared as *Reminiscences of Three Campaigns* in 1919. In his journals relating to the military campaigns he participated in, Ogston repeatedly mentions his impressions will be recorded in a work referred to as 'Scattered Recollections'. It may have been the case, therefore, Ogston projected a larger volume of autobiography, but ultimately opted to only publish material relating to his military campaigns. Ogston's son Walter, editor of the memorial volume *Alexander Ogston K.C.V.O.* – which contains portions of 'Scattered Recollections' and reminiscences of Ogston from colleagues and family – notes Ogston's account of his student days on the continent was written 'after his retirement from the Chair of Surgery' at Aberdeen University in 1898 (*Sir Alexander Ogston K.C.V.O.*, 60). In other sections of the autobiographical material published in *Alexander Ogston K.C.V.O.*, however, Ogston notes the years of composition as being 1919 and 1920. The latter date indicates Ogston was working on a version of 'Scattered Recollections' after the publication of *Reminiscences* – perhaps with a view to a second volume of autobiography, or merely as an account written for his own family.

18. Alexander Ogston, 'Student Days on the Continent', in *Sir Alexander Ogston K.C.V.O.*, 63–85, 64.

19. Ibid. 65.

20. Ibid. 66. Ogston preserved these Lutheran relics in a glass case. His son Walter, however, believed it more than likely Stokes and Ogston had been duped by the museum attendant and that the removed fragments were, in fact, from an ever-replenished stock offered to tourists. See: Walter Ogston, *Sir Alexander Ogston K.C.V.O.*, 66, n. 1.

21. Ibid. 73.

22. Ibid. 82.

23. Ibid. 83.

24. Ibid. 84.

25. Walter Ogston credits this to Ogston's appointment as Ophthalmic Surgeon at Aberdeen Royal Infirmary in 1868.

26. Pennington, *The Modernisation of Medical Teaching at Aberdeen in the Nineteenth Century*, 57.
27. Sylvia Van Kirk, 'Hargrave, James', in *Dictionary of Canadian Biography*, vol. 9. *Dictionary of Canadian Biography* (online).
28. Janet Teissier du Cros, *Cross Currents: A Childhood in Scotland* (East Linton: Tuckwell Press, 1997), 43.
29. Molly Dickens, *A Wealth of Relations: Aberdeen and Shetland* (For Private Circulation, Oxford – MDCLXXII) (Aberdeen University Special Collections Centre (MS 2478/9), 29.
30. du Cros, *Cross Currents*, 43.
31. Dickens, *A Wealth of Relations*, 30.
32. 'Testimonials in Favour of Alexander Ogston, M.D.', Wellcome Trust Collection (online), 16.
33. Ibid. 5.
34. Pennington, *The Modernisation of Medical Teaching at Aberdeen in the Nineteenth Century*, 57.
35. Alexander Ogston, 'Medical training in Aberdeen and in the Scottish universities: an address delivered before the Aberdeen Medical Students' Society on Friday 16th November, 1877' (Aberdeen: Medical Students' Society, 1877), Wellcome Trust Collection (online), 7.
36. Ibid. 9.
37. Ibid. 5.
38. Peter Jones, *A Surgical Revolution Surgery in Scotland 1837 to 1901* (Edinburgh: Birlinn, 2007), 24.
39. Ibid. 24–37.
40. Alexander Ogston, 'How Anaesthetics Came to Aberdeen', in *Sir Alexander Ogston K.C.V.O.*, 92.
41. Alexander Ogston, 'How Antiseptic Surgery Came to Aberdeen', in *Sir Alexander Ogston K.C.V.O.*, 93–7, 95.
42. Ibid. 96.
43. Pennington, *The Modernisation of Medical Teaching at Aberdeen in the Nineteenth Century*, 58.
44. Ogston, 'How Antiseptic Surgery Came to Aberdeen', in *Sir Alexander Ogston K.C.V.O.*, 93–7, 96.
45. Alexander Ogston, 'On the Comparative Strength of Arteries Secured by the Methods of Ligature, Acupressure, and Torsion', *The Lancet*, 93(2381) (1869), 524–6, 525.
46. Alexander Ogston, 'Letter of application for the post of senior surgeon at Aberdeen Royal Infirmary' (1870), Wellcome Trust Collection (online).
47. Alexander Ogston, 'On a new operation for removal of posterior adhesion of the iris' (Aberdeen: Arthur King & Co., 1870). Alexander Ogston, 'On some forms of sudden death, and sudden death in general' (Publisher not identified). Alexander Ogston 'On spontaneous combustion' (Publisher not identified). The other five publications were collected in Alexander Ogston's

'Contributions to Medical Science' (Aberdeen: Arthur King & Co., 1869). All available at Wellcome Trust Collection (online).

48. Alexander Ogston, 'How Antiseptic Surgery Came to Aberdeen', 94.
49. Ibid. 93.
50. Jones, *A Surgical Revolution Surgery in Scotland 1837 to 1901*, 43.
51. Ogston, 'How Antiseptic Surgery Came to Aberdeen' in *Sir Alexander Ogston K.C.V.O.*, 93–7, 93, 94.
52. Jones, *A Surgical Revolution Surgery in Scotland 1837 to 1901*, 121.
53. Alexander Ogston, 'How Antiseptic Surgery Came to Aberdeen', in *Sir Alexander Ogston K.C.V.O.*, 93–7, 94.
54. Ibid. 93, 94.
55. Joseph Lister, 'On the Antiseptic Principle in the Practice of Surgery', *The British Medical Journal*, 2(351) (1867), 246–8, 246. JSTOR.
56. Ibid. 248.
57. Qtd in Peter Jones, *A Surgical Revolution Surgery in Scotland 1837 to 1901*, 108.
58. Ogston, 'How Antiseptic Surgery Came to Aberdeen', in *Sir Alexander Ogston K.C.V.O.*, 93–7, 95.
59. Ibid. 95.
60. Qtd in 'From Dr W. Clark Souter, Aberdeen', in *Sir Alexander Ogston K.C.V.O.*, 163–7, 164.
61. Qtd in Pennington, *The Modernisation of Medical Teaching at Aberdeen in the Nineteenth Century*, 61.
62. Jones, *A Surgical Revolution Surgery in Scotland 1837 to 1901*, 156.
63. 'Sir Henry Gray, K.B.E., C.B., C.M.G., LL.D., F.R.C.S. Ed.', *The British Medical Journal*, 2(4058) (1938), 814–15. JSTOR.
64. 'Extension of the Association in Scotland', *The British Medical Journal*, 1(581) (1872), 189. JSTOR.
65. 'The First Scottish Branch of The British Medical Association', *The British Medical Journal*, 1(587) (1872), 344. JSTOR.
66. Pennington, *The Modernisation of Medical Teaching at Aberdeen in the Nineteenth Century*, 57.
67. du Cros, *Cross Currents: A Childhood in Scotland*, 44.
68. Dickens, *A Wealth of Relations: Aberdeen and Shetland*, 30.
69. du Cros, *Cross Currents: A Childhood in Scotland*, 43, 44.
70. Dickens, *A Wealth of Relations: Aberdeen and Shetland*, 30.
71. Ibid. 33.
72. Walter Henry Ogston, *My Memoirs, Volume 1: 1873–1891*, Aberdeen University Special Collections (MS 3850/4/6/7/3/1), 1.
73. Ibid. 5.
74. Dickens, *A Wealth of Relations: Aberdeen and Shetland*, 37.
75. Ibid. 37–8.
76. Walter Henry Ogston, *My Memoirs, Volume 1: 1873–1891*, 19.
77. Ibid. 4.

78. Ibid. 19.
79. Ibid. 2–3.
80. Ibid. 3–5.

CHAPTER 2

1. Alexander Ogston, 'Discovery of the Germ of Acute Suppuration', in *Sir Alexander Ogston K.C.V.O.*, ed. by Walter Ogston (Aberdeen: Aberdeen University Press, 1943), 98–101, 98.

2. Matthews was a partner, alongside Alexander Marshal Mackenzie, in the architectural firm Matthews & McKenzie. The Directory of Scottish Architects lists both men as being architects of Glendavan, indicating the work was carried out jointly by their firm. 'Glendavan House', DSA Building/Design Report, http://www.scottisharchitects.org.uk/building_full. php?id=209807, accessed 23 February 2022.

3. Molly Dickens, *A Wealth of Relations* (For Private Circulation, Oxford – MDCLXXII) Aberdeen University Special Collections, MS 2478/9, 33.

4. Walter Ogston, 'My Memoirs, Volume 1: 1873–1891', Aberdeen University Special Collections (MS 3850/4/6/7/3/1), 11.

5. Janet Teissier du Cros, *Cross Currents: A Childhood in Scotland* (East Linton: Tuckwell Press, 1997), 44.

6. Walter Ogston, 'Introduction', in *Sir Alexander Ogston K.C.V.O.*, 1–53, 6.

7. Ibid.

8. Alexander Ogston, 'On the Origin of Cancer', Wellcome Trust Collection (online).

9. Alexander Ogston, 'Congenital Malformations of the Lower Jaw', Wellcome Trust Collection (online).

10. Alexander Ogston, 'On Oblique Fracture of the Head of The Humerus', *The Lancet*, 107(2742) (1876), 419–20, Science Direct (online).

11. Peter F. Jones, *A Surgical Revolution: Surgery in Scotland, 1837–1901* (Edinburgh: John Donald, 2007), 124.

12. Alexander Ogston, 'The Operative Treatment of Genu Valgum', *Edinburgh Medical Journal*, March 22(9) (1877), 782–4.

13. Jones, *A Surgical Revolution*, 136.

14. Alexander Ogston, 'The Growth and Maintenance of the Articular Ends of Adult Bones', *Journal of Anatomy and Physiology*, 12(Pt 4) (1878), 503–17, 503–4, National Library of Medicine (online). Here Ogston is summarising the findings of his 1875 paper 'On Articular Cartilage', *Journal of Anatomy and Physiology*, 10(Pt 1), 1–74, National Library of Medicine (online).

15. Ogston, 'The Growth and Maintenance of the Articular Ends of Adult Bones', 505.

16. Peter G. Bullough, *Orthopaedic Pathology*, 5th edn (Maryland Heights, MI: Mosby, 2010), 231.

17. Alexander Ogston, 'An Improved Method of Treating Club-Foot', *Edinburgh Medical Journal*, December 24(6) (1878), 481–92, 481.
18. Ibid. 482. (Emphasis in original.)
19. Ibid. 488.
20. Alexander Ogston, 'Autobiographical Writings', in *Sir Alexander Ogston K.C.V.O.*, 53–110, 98.
21. George Smith, 'Ogston the Bacteriologist', in *The Staphylococci: Proceedings of the Alexander Ogston Centennial Conference*, ed. by Alexander Macdonald and George Smith (Aberdeen: Aberdeen University Press, 1981), 9–21, 11.
22. Olga Amsterdamska, 'Bacteriology, Historical', *International Encyclopedia of Public Health, Volume One*, 2nd edn, ed. by William C. Cockerham (Oxford: Elsevier, 2017), 206–9, 206.
23. Ibid. 206.
24. Smith, 'Ogston the Bacteriologist', 11.
25. Ogston, 'Discovery of the Germ of Acute Suppuration', 98.
26. Smith, 'Ogston the Bacteriologist', 12–13.
27. Theodore Koch, quoted in Smith, 'Ogston the Bacteriologist', 13.
28. Ogston, 'Discovery of the Germ of Acute Suppuration', 98.
29. Ibid. 98.
30. 'Report Upon Micro-Organisms in Surgical Diseases', *The British Medical Journal*, 1(1054) (March 1881), 369–75, 370, JSTOR.
31. Ogston, 'Discovery of the Germ of Acute Suppuration', 99.
32. Ibid.
33. Alexander Ogston, 'Ueber Abscesse' ('On Abscesses', translation by Professor W. Witte), *The Staphylococci: Proceedings of the Alexander Ogston Centennial Conference* (1880), 277–86, 281.
34. Ibid. 285.
35. Ibid. 279.
36. Ibid. 285.
37. Alexander Ogston, 'Journal of visit to Greece, 1881' (7 April 1881–1 May 1881), University of Aberdeen Museums and Special Collections (MS 3850/1/1), 5.
38. Ibid. 7–8.
39. Ibid. 9–10.
40. Ibid. 56.
41. Ibid. 57.
42. Ibid. 72–3.
43. Ibid. 77.
44. Ibid. 109.
45. Ibid. 148–9.
46. Ogston, 'Discovery of the Germ of Acute Suppuration', 100.
47. Ibid. 101.
48. Ibid. 100.

49. Ibid.
50. Joseph Lister, 'An Address on the Relation of Micro-organisms to Inflammation', *The Lancet*, 118(3034) (1881), 696–8, 695, National Library of Medicine (online).
51. Ibid. 696.
52. Ibid. 697.
53. Alexander Ogston, 'Micrococcus Poisoning' (Part One), *Journal of Anatomy and Physiology*, 16(Pt 4) (1882), 526–67, 526, National Library of Medicine (online).
54. Ibid. 532.
55. Ibid. 534.
56. Ibid. 534.
57. Ibid. 561. (Emphasis in original.)
58. Ibid. 562.
59. Ibid. 564.
60. David Grove, *Tapeworms, Lice, and Prions: A Compendium of Unpleasant Infections* (Oxford: Oxford University Press, 2014), 233.
61. Alexander Ogston, 'Micrococcus Poisoning' (Part Two), *Journal of Anatomy and Physiology*, 17(Pt 1) (1882), 24–58, 44, National Library of Medicine (online).
62. Amsterdamska, 'Bacteriology, Historical', 284.
63. Alexander Ogston, 'How Antiseptic Surgery Came to Aberdeen', in *Sir Alexander Ogston K.C.V.O.*, 93–7, 97.
64. Fritz Linder, 'Alexander Ogston', in *The Staphylococci: Proceedings of the Alexander Ogston Centennial Conference*, 1–9, 1.
65. Alan Lyell, 'Alexander Ogston, micrococci, and Joseph Lister', *Journal of the American Academy of Dermatology*, 20(2 Pt 1) (1989), 302–10, 310.
66. Ogston, 'Discovery of the Germ of Acute Suppuration', 101.
67. Ibid. 100.
68. L. R. Hill, 'Taxonomy of the Staphylococci', *The Staphylococci: Proceedings of the Alexander Ogston Centennial Conference*, 33–62, 34.
69. Grove, *Tapeworms, Lice, and Prions*, 234.
70. D. A. Aldeen, and K. Hiramatsu, 'Preface', in *Staphylococcus Aureus: Molecular and Clinical Aspects*, ed. by D. A. Aldeen and K. Hiramatsu (Chichester: Horwood, 2004), xiii–xiv, xiii.
71. Dan Zuberi, *Cleaning Up: How Hospital Outsourcing Is Hurting Workers and Endangering Patients* (Ithaca: ILR Press, 2013), 21.
72. J. R. Adalbert, K. Varshney, R. Tobin, et al., 'Clinical outcomes in patients co-infected with COVID-19 and Staphylococcus aureus: a scoping review', *BMC Infectious Diseases*, 21(985) (2021), 1–17, 1.
73. Ibid. 15.
74. Alexander Ogston, 'Journal of visit to Norway, 1882' (21 July 1882–1 September 1882), University of Aberdeen Museums and Special Collections (MS 3850/1/2), 19.

75. Ibid. 145.
76. Ibid. 55–6.
77. Ibid. 144, 146.

CHAPTER 3

1. Alexander Ogston, *Reminiscences of Three Campaigns* (London: Hodder & Stoughton, 1919), 1.
2. Alexander Ogston, 'On Flat-Foot and Its Cure by Operation', *Bristol Medico-Chirurgical Journal*, 2(3) (1884), 1–20. National Library of Medicine (online).
3. K. Muntarbhorn and S. Thanaviratananich, 'Mini-anterior and Combined Frontal Sinusotomy and Drilling of the Nasofrontal Beak', in *Micro-endoscopic Surgery of the Paranasal Sinuses and the Skull Base*, ed. by A. C. Stamm and W. Draf (Berlin: Springer, 2000), 279–86, 279.
4. Friedrich Busch, *Handbook of General Therapeutics*, vol. 5, ed. by Hugo Ziemssen (New York: William Wood, 1886), 65.
5. 'Scotland', *The British Medical Journal*, 1(1262) (1885), 502. JSTOR.
6. Carolyn Pennington, *The Modernisation of Medical Teaching at Aberdeen in the 19th Century* (Aberdeen: Aberdeen University Press, 1994), 66.
7. Qtd in Ibid. 67.
8. Ogston, *Reminiscences of Three Campaigns*, 8.
9. Ibid. 11. (Emphasis in original.)
10. Ibid. 12–13.
11. Ibid. 16.
12. Ibid. 17.
13. Ibid. 19–20.
14. Ibid. 20.
15. Ibid.
16. Ibid. 24.
17. Ibid. 42.
18. *The British Medical Journal: Volume II for 1885, July to December*, ed. by Ernest Hart (London: British Medical Association, 1885), 845.
19. John McConachie, *The Student Soldiers* (Elgin: Moravian Press, 1995), 5.
20. Pennington, *The Modernisation of Medical Teaching at Aberdeen in the 19th Century*, 68.
21. Walter Ogston, 'Introduction', in *Sir Alexander Ogston K.C.V.O.*, ed. by Walter Ogston (Aberdeen: Aberdeen University Press, 1943), 1–52, 13.
22. Ibid. 39–40.
23. Molly Dickens, *A Wealth of Relations: Aberdeen and Shetland* (For Private Circulation, Oxford – MDCLXXII) (Aberdeen University Special Collections, MS 2478/9), 65.
24. 'Sir Alexander Ogston', *The British Medical Journal*, 1(1634) (1892), 875. JSTOR.

25. Sir Alexander Ogston, qtd in 'Scotland', *The British Medical Journal*, 1(1636) (1892), 985. JSTOR.

26. Pennington, *The Modernisation of Medical Teaching at Aberdeen in the 19th Century*, 69.

27. 'Scotland', *The British Medical Journal*, 1(1638) (1892), 1100–01, 1101. JSTOR.

28. Alexander Ogston, 'An Audience of Queen Victoria', in *Sir Alexander Ogston K.C.V.O.*, 101–10, 101.

29. Ibid. 102.

30. Ibid. 103.

31. Ibid. 104–5.

32. Ibid. 106.

33. Ibid.

34. Ibid. 107.

35. Ibid.

36. Queen Victoria, 'Journal Entry: Monday 26th September 1892', Queen Victoria's Journals, 96 (1st August 1892–31st December 1892), http://www.queenvictoriasjournals.org/search/displayItem.do?FormatType=full textimgsrc&QueryType=articles&ResultsID=3305436364666&filter Sequence=0&PageNumber=1&ItemNumber=1&ItemID=qvj21776& volumeType=PSBEA, accessed 16 May 2022.

37. Alexander Ogston, 'An Audience of Queen Victoria', in *Sir Alexander Ogston K.C.V.O.*, 109–110.

38. Ibid. 108.

39. Walter Ogston, 'Introduction', in *Sir Alexander Ogston K.C.V.O.*, 21.

40. Janet Teissier du Cros, *Cross Currents: A Childhood in Scotland* (East Linton: Tuckwell Press, 1997), 59.

41. Ogston, 'Sketches of visit to South Africa, 1894–1895/Journal of visit to Russia, 1898–1899', University of Aberdeen Museums and Special Collections (MS 3850/1/4), 38.

42. Alexander Ogston, 'Boer War Journal (volume 1), 1899–1900', University of Aberdeen Museums and Special Collections (MS 3850/1/5), 189.

43. Herbert Grierson, *Vita Mea: The Autobiography of Sir Herbert Grierson*, ed. by Cairns Craig (Aberdeen: Aberdeen University Press, 2014), 85.

44. Ibid. 88.

45. Ogston, *Reminiscences of Three Campaigns*, 43.

46. Queen Victoria, qtd in Peter Lovegrove's *Not Least in the Crusade: A Short History of the Royal Army Medical Corps* (Aldershot: Gale & Polden, 1951), 13.

47. Lovegrove, *Not Least in the Crusade*, 9.

48. Tom Scotland, and Steven Heys, *Wars, Pestilence, and the Surgeon's Blade* (Solihull: Helion, 2013), 254.

49. Lovegrove, *Not Least in the Crusade*, 15–16.

50. Scotland and Heys, *Wars, Pestilence, and the Surgeon's Blade*, 271.

51. Ibid. 271.
52. Alexander Ogston, 'Sketches of visit to South Africa, 1894–1895/Journal of visit to Russia, 1898–1899', 5.
53. Ibid. 7.
54. Ibid. 8.
55. Ibid. 29.
56. Ibid. 32.
57. Ibid. 36.
58. Alexander Ogston, 'Address in Surgery: The Medical Services of The Army and Navy', *The British Medical Journal*, 2(2014) (1899), 337–45, 338. JSTOR.
59. Ibid. (Ellipsis in original.)
60. Ibid. 339.
61. Ibid. 340.
62. Ibid. 341.
63. Ibid. 342.
64. Ibid. 343.
65. Ibid.
66. Ibid. 344.
67. Ibid.
68. Ibid. 345.
69. 'Sixty-Seventh Annual Meeting of The British Medical Association', *The British Medical Journal*, 2(2015) (1899), 429–34, 429. JSTOR.
70. 'Royal Navy and Army Medical Services', *The British Medical Journal*, 2(2021) (1899), 817–19, 817. JSTOR.
71. Ogston, *Reminiscences of Three Campaigns*, 49. (Emphasis in original.)

CHAPTER 4

1. Arthur Bigge, qtd in Alexander Ogston, *Reminiscences of Three Campaigns* (London: Hodder & Stoughton, 1919), 49.
2. Nicholas Murray, *The Rocky Road to the Great War: The Evolution of Trench Warfare to 1914* (Washington, DC: Potomac Books, 2013), 81–90.
3. William Boothby, *Weapons and the Law of Armed Conflict* (Oxford: Oxford University Press, 2016), 135.
4. Harold E. Raugh, 'Bullets' in *The Victorians at War, 1815–1914: An Encyclopaedia of British Military History* (Santa Barbara, CA: ABC-CLIO, 2004), 64–5.
5. Ibid. 65.
6. Alexander Ogston, 'The Effects of the Dum-dum Bullet from a Surgical Point of View', *The British Medical Journal*, 1(1952) (May 1898), 1425. National Library of Medicine (online).
7. Alexander Ogston, 'The Peace Conference and The Dum-dum Bullet', *The British Medical Journal*, 2(2013) (1899), 278–81, 279. JSTOR.

8. Ogston, 'The Effects of the Dum-dum Bullet from a Surgical Point of View', 1425.

9. Alexander Ogston, 'The Wounds Produced by Modern Small-Bore Bullets: The Dum-dum Bullet and the Soft-Nosed Mauser', *The British Medical Journal*, 2(1968) (September 1898), 813–15, 814. National Library of Medicine (online).

10. Alexander Ogston, 'Continental Criticism of English Rifle Bullets', *The British Medical Journal*, 1(1995) (1899), 752–7, 756. JSTOR.

11. Alexander Ogston, 'Address in Surgery: The Medical Services of The Army and Navy', *The British Medical Journal*, 2(2014) (1899), 337–45, 338. JSTOR.

12. 'Declaration (IV,3) concerning Expanding Bullets. The Hague, 29 July 1899', International Committee of the Red Cross, https://ihl-databases. icrc.org/applic/ihl/ihl.nsf/Article.xsp?action=openDocument&document Id=F5FF4D9CA7E41925C12563CD0051616B, accessed 18 July 2022.

13. Ogston, 'The Peace Conference and The Dum-dum Bullet', 279.

14. Ibid. 280.

15. Ogston, 'The Peace Conference and The Dum-dum Bullet', 280.

16. Ibid. 280–1.

17. Ibid. 281.

18. Stephen M. Miller, *Lord Methuen and the British Army: Failure and Redemption in South Africa* (London: Frank Cass, 1999), 115.

19. Arthur Bigge, qtd in Ogston, *Reminiscences of Three Campaigns*, 49.

20. Ibid. 51–2.

21. Alexander Ogston, 'Boer War Journal (volume 1), 1899–1900', University of Aberdeen Museums and Special Collections (MS 3850/1/5), 8.

22. Ibid. 6.

23. Ogston, *Reminiscences of Three Campaigns*, 62.

24. Ibid. 71.

25. Ibid. 75.

26. Ibid. 79.

27. Ibid. 81.

28. Ibid. 83.

29. Ogston, 'Boer War Journal (volume 1), 1899–1900', 61.

30. Ibid. 77.

31. Ibid. 78.

32. Ibid. 87.

33. Ogston, *Reminiscences of Three Campaigns*, 99.

34. Ogston, 'Boer War Journal (volume 1), 1899–1900', 92.

35. Ogston, *Reminiscences of Three Campaigns*, 106.

36. Ogston, 'Boer War Journal (volume 1), 1899–1900', 106.

37. Ogston, *Reminiscences of Three Campaigns*, 108.

38. Ibid. 108–9.

39. Ogston, 'Boer War Journal (volume 1), 1899–1900', 106.

40. Ogston, *Reminiscences of Three Campaigns*, 117.
41. Ibid. 118.
42. Ibid. 119.
43. Ibid. 119.
44. Wilfred Owen, 'Dulce Et Decorum Est', in *The Poems of Wilfred Owen*, ed. by Edmund Blunden (1946; rep. London: Chatto & Windus), 66.
45. Ibid. 119.
46. Ibid. 125.
47. Ogston, 'Boer War Journal (volume 1), 1899–1900', 128.
48. Ibid. 129.
49. Ibid. 130.
50. Ibid. 132.
51. Ibid. 133.
52. Ogston, *Reminiscences of Three Campaigns*, 133.
53. Ibid. 134.
54. Ibid. 135.
55. Ibid. 137.
56. Ibid. 139.
57. Ibid. 141.
58. Ogston, 'Boer War Journal (volume 1), 1899–1900', 138.
59. Ibid. 153.
60. Ibid. 154.
61. Ibid. 155.
62. Ibid. 155.
63. Ibid. 156.
64. Ibid. 157.
65. Ibid.
66. Brigadier-General Charles Ogston, 'From Brigadier-General Charles Ogston', in *Sir Alexander Ogston K.C.V.O.*, ed. by Walter Ogston (Aberdeen: Aberdeen University Press, 1943), 137–9, 137.
67. Ogston, *Reminiscences of Three Campaigns*, 149.
68. Ogston, 'Boer War Journal (volume 1), 1899–1900', 174.
69. Ibid. 174, 175.
70. Ibid. 178.
71. Alexander Ogston, 'Boer War Journal (volume 2), 1900', University of Aberdeen Museums and Special Collections (MS 3850/1/6), 8.
72. Ibid. 11–12.
73. Ibid. 14.
74. Ibid. 19.
75. Ibid. 31.
76. Ibid. 31–2.
77. Ibid. 35.
78. Ibid. 40.
79. Ibid. 45.

80. Ibid. 45.
81. Ibid. 48.
82. Ibid. 61–2.
83. Ibid. 63.
84. Ibid. 63.
85. Ibid. 68.
86. Ibid. 74.
87. Ibid. 76.
88. Ibid. 77.
89. Ibid. 78.
90. Ibid. 80.
91. Ibid. 81.
92. Ibid.
93. Ibid.
94. Ibid. 83.
95. Ibid. 89.
96. Ibid. 93.
97. Ibid. 97.
98. Ibid. 98.
99. Ibid. 99.
100. Ibid. 104, 106.
101. Ibid. 106.
102. Ibid. 107.
103. Ibid. 108.
104. Ibid. 107.
105. See: Fred R. Van Hartesveldt, *The Boer War: Historiography and Annotated Bibliography* (Westport, CT: Greenwood Press, 2000), 32; Howard C. Hillegas, *With the Boer Forces* (Frankfurt: Outlook Verlag, 2020), 138.
106. Van Hartesveldt, *The Boer War: Historiography and Annotated Bibliography*, 33.
107. Ogston, 'Boer War Journal (volume 1), 1899–1900', 15.
108. Ibid. 152.
109. Ibid. 160.
110. Ogston, 'Boer War Journal (volume 2), 1900', 10.
111. Ibid. 152.
112. Ibid. 44.
113. Ibid. 84.
114. Ibid. 112.
115. This inscription was added (on the same memorial stone) beneath one dedicated to Hargrave's grandfather (Ogston's father) Francis Ogston in St. Nicholas Church, Aberdeen. An identical memorial stone (which is solely dedicated to Sir Alexander Ogston) is positioned to the left of the Francis/Hargrave memorial in Drum's Aisle. https://www.findagrave.com/memorial/153841270/francis-ogston, accessed 18 July 2022.

CHAPTER 5

1. 'Aberdeen, July, 1914', *The British Medical Journal*, 2(2740) (1913), 37. JSTOR.
2. 'Royal Navy and Army Medical Services', *The British Medical Journal*, 1(2095) (1901), 492–3. JSTOR.
3. Peter Lovegrove, *Not Least in the Crusade: A Short History of the Royal Army Medical Corps* (Aldershot: Gale & Polden, 1951), 26.
4. 'The R.A.M.C. Expert Committee', *The British Medical Journal*, 2(2114) (1901), 31–3. JSTOR.
5. 'Personal Record', in *Sir Alexander Ogston K.C.V.O.*, ed. by Walter Ogston (Aberdeen: Aberdeen University Press, 1943), 53–62, 59.
6. 'Report of the Royal Commission on the South African War', *The British Medical Journal*, 2(2226) (1903), 484–7. JSTOR.
7. Ibid. 486.
8. Lovegrove, *Not Least in the Crusade*, 32.
9. Alexander Ogston, 'A New Principle of Curing Club-Foot in Severe Cases in Children a Few Years Old', *The British Medical Journal*, 1(2164) (1902), 1524–6, 1524. JSTOR.
10. *Report of the Royal Commission on Physical Training (Scotland), Vol. 1.* (Edinburgh: His Majesty's Stationery Office, 1903), 7.
11. 'The Royal Commission on Physical Training (Scotland)', *The British Medical Journal*, 1(2205) (1903), 817. JSTOR.
12. Elizabeth Crawford, *The Women's Suffrage Movement: A Reference Guide 1866–1928* (London: Taylor & Francis, 2003), 472.
13. Qtd in Susan Kingsley Kent, *Sex and Suffrage in Britain, 1860–1914* (Princeton University Press, 2014), 173.
14. 'Unstoppable Voices: The Woman with the Whip', https://www.royalalberthall.com/about-the-hall/news/2021/march/unstoppable-voices-the-woman-with-the-whip/, accessed 31 July 2022.
15. E. Sylvia Pankhurst, *The Suffragette: The History of the Women's Militant Suffrage Movement, 1905–1910* (New York: Sturgis & Walton, 1911), 346–7.
16. Walter Ogston, 'Introduction', in *Sir Alexander Ogston K.C.V.O.*, 1–52, 47.
17. Helen Ogston, 'From Helen C.E.D. Ogston', in *Sir Alexander Ogston K.C.V.O.*, 111.
18. W. Douglas Simpson, 'Editor's Preface', *The Prehistoric Antiquities of the Howe of Cromar* by Sir Alexander Ogston (Aberdeen: Aberdeen University Press, 1931), v–vii, v. As Simpson outlines in his 'Preface', Ogston's manuscript comprised three main sections: a lengthy review of relevant extant literature, a systematic description of the remains in the Howe, and a third part where Ogston drew parallels between the remains he had studied and those found elsewhere. *The Prehistoric Antiquities* published in 1931 presents only the second portion of Ogston's manuscript.

19. Alexander Ogston, 'Author's Preface', *The Prehistoric Antiquities of the Howe of Cromar*, ix–xiii, ix.
20. Alexander Ogston, 'An Audience of Queen Victoria', in *Sir Alexander Ogston K.C.V.O.*, 101–110, 110.
21. Walter Ogston, 'Introduction', in *Sir Alexander Ogston K.C.V.O.*, 1–52, 45.
22. Ibid. 34–5.
23. 'Aberdeen, July, 1914', 37. JSTOR.

CHAPTER 6

1. Alexander Ogston, *Reminiscences of Three Campaigns* (London: Hodder & Stoughton, 1919), 298.
2. 'Aberdeen, July, 1914', *The British Medical Journal*, 2(2740) (1913), 37. JSTOR.
3. 'Annual Meeting, Aberdeen', *The British Medical Journal*, 1(2790) (1914), 1375. JSTOR.
4. 'Sir Alexander Ogston', *The British Medical Journal*, 1(2790) (1914), 1375. JSTOR.
5. 'Golf Courses at Aberdeen and in its Vicinity', *The British Medical Journal*, 2(2792) (1914), 9–10. JSTOR.
6. 'The President's Address', *The British Medical Journal*, 2(2796) (1914), 253–54. JSTOR.
7. 'Annual General Meeting', *The British Medical Journal*, 2(2796) (1914), 128–31, 129. JSTOR.
8. 'Annual General Meeting', 130.
9. *Aberdeen University Review, Vol. II* (Aberdeen: Aberdeen University Press, 1915), 85.
10. 'Eighty-Second Annual Meeting of The British Medical Association', *The British Medical Journal*, 2(2797) (1914), 137–42, 140. JSTOR.
11. Arthur Anderson Martin, *A Surgeon in Khaki* (London: Edward Arnold, 1915), 2–3.
12. 'The Curtain', *The British Medical Journal*, 2(2797) (1914), 308–9. JSTOR.
13. Alexander Ogston, 'Letter to Sir James Reid, 3rd August, 1914', Aberdeen Medico-Chirurgical Society (AMCS/4/10/5/1).
14. George V, 'Letter to Sir James Reid, 6th September, 1914', Aberdeen Medico-Chirurgical Society (AMCS/4/10/5/4).
15. 'Bulletin signed by John Marnoch, James Reid and Alexander Ogston, 9th September, 1914', Aberdeen Medico-Chirurgical Society (AMCS/4/10/5/12).
16. 'Voluntary Aid Detachments. An Auxiliary War Hospital', *The British Medical Journal*, 1(2818) (1915), 38–9. JSTOR.

17. James A. Davidson, 'From Dr. James A. Davidson', in *Sir Alexander Ogston K.C.V.O.*, ed. by Walter Ogston (Aberdeen: Aberdeen University Press, 1943), 143–5, 144.
18. E. A. Chill, qtd in *Sir Alexander Ogston K.C.V.O.*, 179.
19. Alexander Ogston, 'First World War Journal (volume 1), 1915–1916', University of Aberdeen Museums and Special Collections (MS 3850/1/7), 1.
20. Ogston, *Reminiscences of Three Campaigns*, 236–7.
21. Ibid. 237.
22. Ibid. 238.
23. Ibid.
24. Alexander Ogston, 'First World War Journal (volume 1), 1915–1916', University of Aberdeen Museums and Special Collections (MS 3850/1/7), 16.
25. Ibid. 18.
26. Ibid. 17
27. Ibid. 16.
28. Ibid. 17.
29. Ibid. 29–30.
30. Ibid. 32.
31. Ibid. 38.
32. *Sir Alexander Ogston K.C.V.O.*, 145, 169.
33. Alexander Ogston, 'First World War Journal (volume 1), 1915–1916', 38.
34. 'Annual Report of Council, 1914–15', *The British Medical Journal*, 1(2836) (1915), 166–237, 166. JSTOR.
35. Ogston, 'First World War Journal (volume 1), 1915–1916', 41.
36. Ogston, *Reminiscences of Three Campaigns*, 254.
37. Ogston, 'First World War Journal (volume 1), 1915–1916', 56.
38. Ibid. 58.
39. Ibid. 59.
40. Ibid. 60.
41. Ibid. 69.
42. Ogston, 'First World War Journal (volume 1), 1915–1916', 67–70.
43. Ogston, *Reminiscences of Three Campaigns*, 259.
44. 'The Mustering of The Profession. Medical Service in The Great War, 1914–19', *The British Medical Journal*, 1(3236) (1923), 24–9, 25. JSTOR.
45. 'The War Emergency', *The British Medical Journal*, 2(2857) (1915), 147–8. JSTOR.
46. 'Scotland', *The British Medical Journal*, 2(2859) (1915), 587. JSTOR.
47. Ogston, 'First World War Journal (volume 1), 1915–1916', 88.
48. Alexander Ogston, 'Our Wounded – Sphagnum Moss as a Dressing', *The National Review*, 67(402) (August 1916), 870–5, 872. Internet Archive.
49. Ibid. 875.
50. Ogston, 'First World War Journal (volume 1), 1915–1916', 88.
51. Ogston, *Reminiscences of Three Campaigns*, 261–2.

52. Ogston, 'First World War Journal (volume 1), 1915–1916', 100–1.
53. Ibid. 102.
54. Ogston, *Reminiscences of Three Campaigns*, 262.
55. Ogston, 'First World War Journal (volume 1), 1915–1916', 99.
56. George Macaulay Trevelyan, *Scenes from Italy's War* (Boston: Houghton Mifflin Company, 1919), 103–4.
57. Ibid. 107.
58. Freya Stark, *Traveller's Prelude* (London: John Murray, 1950), 180, 183.
59. Ogston, *Reminiscences of Three Campaigns*, 263.
60. Ogston, 'First World War Journal (volume 1), 1915–1916', 103–4.
61. Laura Trevelyan, *A Very British Family: The Trevelyans and Their World* (London: I. B. Tauris & Co. Ltd, 2006), 150.
62. Edward Verrall Lucas, *Outposts of Mercy: The record of a visit in November and December 1916, to the various units of the British Red Cross in Italy* (London: Methuen & Co., 1916), 29.
63. Ogston, *Reminiscences of Three Campaigns*, 268.
64. Ibid. 267.
65. Ogston, *Reminiscences of Three Campaigns*, 183–4.
66. *Reports by the Joint War Committee and the Joint Finance Committee of the British Red Cross Society and the Order of St. John of Jerusalem in England on Voluntary Aid rendered to the Sick and Wounded at Home and Abroad and to British Prisoners of War, 1914–1919, with appendices* (London: His Majesty's Stationery Office, 1921), 434.
67. Ogston, *Reminiscences of Three Campaigns*, 290.
68. Ogston, 'First World War Journal (volume 1), 1915–1916', 117.
69. 'Sir A. Ogston's Appeal for Ambulances for Italy.' Single sheet of typed text, preserved in Sir Alexander Ogston's 'First World War Journal (volume 1), 1915–1916', University of Aberdeen Museums and Special Collections (MS 3850/1/7).
70. Ibid. 124.
71. Ogston, 'First World War Journal (volume 1), 1915–1916', 120.
72. Ibid. 125.
73. G. M. Trevelyan, 'A Memorandum on the Position of the B.R.C.S. Unit, Italy.' Loose printed sheets (4 pages) preserved within Sir Alexander Ogston, 'First World War Journal (volume 1), 1915–1916', University of Aberdeen Museums and Special Collections (MS 3850/1/7).
74. Sir Alexander Ogston, 'Memorandum by Sir Alexander Ogston K.C.V.O., MD.' Loose printed sheets (3 pages) preserved within Sir Alexander Ogston, 'First World War Journal (volume 1), 1915–1916', University of Aberdeen Museums and Special Collections (MS 3850/1/7).
75. Ogston, 'First World War Journal (volume 1), 1915–1916', 126–7.
76. Ibid. 138.
77. Ibid. 126.
78. Ibid. 127.

79. Ibid. 129.

80. Ibid. 137.

81. Ibid. 145–6.

82. Ibid. 140.

83. Ibid. 147.

84. Ibid. 147–8.

85. Ibid. 148–9.

86. Ibid. 149.

87. Ibid. 150.

88. Ibid. 157.

89. Alexander Ogston, 'First World War Journal (volume 2), 1916–1917', University of Aberdeen Museums and Special Collections (MS 3850/1/8), 159.

90. Ogston, *Reminiscences of Three Campaigns*, 281.

91. Ibid. 287–8.

92. Lucas, *Outposts of Mercy*, 16.

93. Ogston, *Reminiscences of Three Campaigns*, 317.

94. Lucas, *Outposts of Mercy*, 18.

95. Ogston, 'First World War Journal (volume 2), 1916–1917', 170.

96. Ibid. 248.

97. Ibid. 172.

98. Ibid. 176.

99. Ogston, *Reminiscences of Three Campaigns*, 318.

100. Ibid. 319.

101. Ibid. 318.

102. Ogston, 'First World War Journal (volume 2), 1916–1917', 175.

103. Ibid. 177.

104. Ibid. 179.

105. Ibid. 181.

106. Ogston, 'First World War Journal (volume 2), 1916–1917', 189.

107. Ibid. 193.

108. Ibid. 207.

109. Ibid. 211.

110. Ogston, *Reminiscences of Three Campaigns*, 292.

111. Ibid. 320.

112. Ogston, 'First World War Journal (volume 2), 1916–1917', 216.

113. Laura Trevelyan, *A Very British Family*, 165.

114. Ogston, 'First World War Journal (volume 2), 1916–1917', 219.

115. Ibid. 222.

116. Ogston, *Reminiscences of Three Campaigns*, 321.

117. Ogston, 'First World War Journal (volume 2), 1916–1917', 231.

118. Ibid. 260.

119. Ogston, 'First World War Journal (volume 2), 1916–1917', 266.

120. Ibid. 267.

121. Ogston, *Reminiscences of Three Campaigns*, 275.

122. Ogston, 'First World War Journal (volume 2), 1916–1917', 267.

123. Ogston, *Reminiscences of Three Campaigns*, 327.

124. Ibid. 329.

125. Ogston, 'First World War Journal (volume 2), 1916–1917', 289.

126. Alexander Ogston, 'Letter to Bevan B. Baker, 27th November, 1917'. Loose printed sheet preserved within Sir Alexander Ogston, 'First World War Journal (volume 2)', University of Aberdeen Museums and Special Collections (MS 3850/1/8).

127. *Aberdeen University Review*, Vol. V (Aberdeen: Aberdeen University Press, 1918), 270.

128. Herbert Grierson, 'From Sir Herbert Grierson' from *Sir Alexander Ogston K.C.V.O.*, 117–21, 120.

129. Gina Kolata, *The Story of the Great Influenza Pandemic and the Search for the Virus that Caused It* (New York: Simon and Schuster, 1999), 7.

130. Lisa and Kevin Freeman-Cook, *Staphylococcus Infections* (New York: Chelsea House Publishers, 2006), 35.

131. Robert Lewis, 'Hemingway in Italy: Making It Up', *Journal of Modern Literature* 9(2) (May 1982), 209–36; *David* A. Rennie, 'The *Real British Red Cross* and Hemingway's A *Farewell to Arms*', *Hemingway Review* 37(2) (Spring 2018), 25–41.

132. Ernest Hemingway, *A Farewell to Arms* (1929; rep. London: Heinemann, 2012), 15.

133. George Macaulay Trevelyan, *Scenes from Italy's War*, 55.

134. Hemingway, *A Farewell to Arms* 15, 24.

135. Ogston, *Reminiscences of Three Campaigns*, 321

136. Michael Reynolds, *Hemingway's First War: The Making of* A Farewell to Arms (Princeton: Princeton University Press, 1975), 95.

137. Reynolds, *Hemingway's First War*, 100.

138. Hemingway, *A Farewell to Arms*, 22.

139. Ibid. 31.

140. Trevelyan, *Scenes from Italy's War*, 112.

141. Ogston, *Reminiscences of Three Campaigns*, 289.

142. Ibid. 290.

143. Geoffrey Young, *The Grace of Forgetting* (London: Country Life Ltd, 1953), 285.

144. Ogston, *Reminiscences of Three Campaigns*, 290

145. Trevelyan, *Scenes from Italy's War*, 182–3; Hemingway, *A Farewell to Arms*, 173.

146. Trevelyan, *Scenes from Italy's War*, 186; Hemingway, *A Farewell to Arms*, 191–4.

147. Ogston, *Reminiscences of Three Campaigns*, 288–9.

148. Ibid. 298.

CHAPTER 7

1. Herbert Grierson, 'Preface', in *Sir Alexander Ogston K.C.V.O.*, ed. by Walter Ogston (Aberdeen: Aberdeen University Press, 1943), v–x, ix.
2. '[Anonymous review of] Three Campaigns', *The British Medical Journal*, 1(3081) (1920), 87–8, 88. JSTOR.
3. 'Annual Representative Meeting. Friday, July 21st', *The British Medical Journal*, 2(3213) (1922), 41–62, 42. JSTOR.
4. Walter Ogston, 'Introduction', in *Sir Alexander Ogston K.C.V.O.*, 1–52, 34–5.
5. Qtd in Tom Scotland and Ann Boyer's *Henry Gray: Surgeon in the Great War* (Edinburgh: Capercaillie Books, 2015), 60.
6. Qtd in 'Scotland', *The British Medical Journal*, 2(3262) (1923), 41–2, 41. JSTOR.
7. 'Reports of Societies', *The British Medical Journal*, 2(3279) (1923), 811–14, 814. JSTOR.
8. Alexander G. Ogston, 'From Alexander G. Ogston', in *Sir Alexander Ogston K.C.V.O.*, 132–4, 133.
9. Walter Ogston, 'Notes on the Reminiscences', in *Sir Alexander Ogston K.C.V.O.*, 169–71, 169.
10. Mary Grierson, 'From Mary Grierson', in *Sir Alexander Ogston K.C.V.O.*, 134–6, 136.
11. 'From the British Medical Journal', in *Sir Alexander Ogston K.C.V.O.*, 172–80, 175.
12. Ibid. 179.
13. Walter Ogston, 'Epilogue', in *Sir Alexander Ogston K.C.V.O.*, 188–190, 189.
14. Constance Ogston, 'From Constance Ogston', in *Sir Alexander Ogston K.C.V.O.*, 112–16, 116.
15. Herbert Grierson, 'From Herbert Grierson', in *Sir Alexander Ogston K.C.V.O.*, 117–21, 119.
16. Qtd in 'From the British Medical Journal', in *Sir Alexander Ogston K.C.V.O.*, 172–80, 173.
17. Aberdeen Archives, Gallery & Museums, 'Sir Alexander Ogston PLAQUE097', https://emuseum.aberdeencity.gov.uk/sites/264/plaque097, accessed 1 November 2022.

Further Reading

Adalbert, J. R., K. Varshney, R. Tobin, et al. 'Clinical outcomes in patients co-infected with COVID-19 and Staphylococcus aureus: a scoping review', *BMC Infectious Diseases* 21(985) (2021), 1–17

Aldeen, D. A. and K. Hiramatsu, 'Preface', in *Staphylococcus Aureus: Molecular and Clinical Aspects*, ed. by D. A. Aldeen and K. Hiramatsu (Chichester: Horwood, 2004), xiii–xiv

Amsterdamska, Olga, 'Bacteriology, Historical', in *International Encyclopedia of Public Health, Volume One*, 2nd edn, ed. by William C. Cockerham (Oxford: Elsevier, 2017), 206–9

Boothby, William, *Weapons and the Law of Armed Conflict* (Oxford: Oxford University Press, 2016)

Bullough, Peter G., *Orthopaedic Pathology*, 5th edn (Maryland Heights, MI: Mosby, 2010)

Busch, Friedrich, *Handbook of General Therapeutics*, vol. 5, ed. by Hugo Ziemssen (New York: William Wood, 1886)

Crawford, Elizabeth, *The Women's Suffrage Movement: A Reference Guide 1866–1928* (London: Taylor & Francis, 2003)

Dickens, Molly, *A Wealth of Relations: Aberdeen and Shetland* (For Private Circulation, Oxford – MDCLXXII) (University of Aberdeen Museums and Special Collections, MS 2478/9)

du Cros, Janet Teissier, *Cross Currents: A Childhood in Scotland* (East Linton: Tuckwell Press, 1997)

Freeman-Cook, Lisa and Kevin, *Staphylococcus Infections* (New York: Chelsea House Publishers, 2006)

Grierson, Herbert, *Vita Mea: The Autobiography of Sir Herbert Grierson*, ed. by Cairns Craig (Aberdeen: Aberdeen University Press, 2014)

Grove, David, *Tapeworms, Lice, and Prions: A Compendium of Unpleasant Infections* (Oxford: Oxford University Press, 2014)

Hartesveldt, Fred R. Van, *The Boer War: Historiography and Annotated Bibliography* (Westport, CT: Greenwood Press, 2000)

Hemingway, Ernest, *A Farewell to Arms* (1929; rep. London: Heinemann, 2012)

Hillegas, Howard C., *With the Boer Forces* (Frankfurt: Outlook Verlag, 2020)

Jones, Peter, *A Surgical Revolution Surgery in Scotland 1837 to 1901* (Edinburgh: Birlinn, 2007)

Kolata, Gina, *The Story of the Great Influenza Pandemic and the Search for the Virus that Caused It* (New York: Simon and Schuster, 1999)

Lewis, Robert, 'Hemingway in Italy: Making It Up', *Journal of Modern Literature*, 9(2) (May 1982), 209–36

Lister, Joseph, 'On the Antiseptic Principle in the Practice of Surgery', *The British Medical Journal*, 2(351) (1867), 246–8

— 'An Address on the Relation of Micro-organisms to Inflammation', *The Lancet*, 118(3034) (1881), 696–8

Lovegrove, Peter, *Not Least in the Crusade: A Short History of the Royal Army Medical Corps* (Aldershot: Gale & Polden, 1951)

Lucas, Verrall Edward, *Outposts of Mercy: The record of a visit in November and December 1916, to the various units of the British Red Cross in Italy* (London: Methuen & Co., 1916)

Lyell, Alan, 'Alexander Ogston, micrococci, and Joseph Lister', *Journal of the American Academy of Dermatology*, 20(2 Pt 1) (1989), 302–10

Macdonald, Alexander and George Smith, eds, *The Staphylococci: Proceedings of the Alexander Ogston Centennial Conference* (Aberdeen: Aberdeen University Press, 1981)

McConachie, John, *The Student Soldiers* (Elgin: Moravian Press, 1995)

Martin, Arthur Anderson, *A Surgeon in Khaki* (London: Edward Arnold, 1915)

Miller, Stephen M., *Lord Methuen and the British Army: Failure and Redemption in South Africa* (London: Frank Cass, 1999)

Muntarbhorn, K. and S. Thanaviratananich, 'Mini-anterior and Combined Frontal Sinusotomy and Drilling of the Nasofrontal Beak', in *Micro-endoscopic Surgery of the Paranasal Sinuses and the Skull Base*, ed. by A. C. Stamm and W. Draf (Berlin: Springer, 2000), 279–286

Murray, Nicholas, *The Rocky Road to the Great War: The Evolution of Trench Warfare to 1914* (Washington, DC: Potomac Books, 2013)

Ogston, Alexander, *A Genealogical History of the Families of Ogston from their First Appearance circa. A.D. 1200* (privately printed, 1876)

— 'Contributions to Medical Science' (Aberdeen: Arthur King & Co., 1869)

— 'On the Comparative Strength of Arteries Secured by the Methods of Ligature, Acupressure, and Torsion', *The Lancet*, 93(2381) (1869), 524–6

— 'On a new operation for removal of posterior adhesion of the iris' (Aberdeen: Arthur King & Co., 1870).

— 'On Oblique Fracture of the Head of The Humerus', *The Lancet*, 107(2742) (1876), 419–20

— 'The Operative Treatment of Genu Valgum', *Edinburgh Medical Journal*, March 22(9) (1877), 782–4

— 'The Growth and Maintenance of the Articular Ends of Adult Bones', *Journal of Anatomy and Physiology*, 12(Pt 4) (1878), 503–17

— 'An Improved Method of Treating Club-Foot', *Edinburgh Medical Journal*, December 24(6) (1878), 481–92

— 'Report Upon Micro-Organisms in Surgical Diseases', *The British Medical Journal*, 1(1054) (March 1881), 369–75

— 'Ueber Abscesse' ('On Abscesses', translation by Professor W. Witte), *The Staphylococci: Proceedings of the Alexander Ogston Centennial Conference* (1880), 277–86

— 'Journal of visit to Greece, 1881' (7 April 1881–1 May 1881), University of Aberdeen Museums and Special Collections (MS 3850/1/1)

— 'Journal of visit to Norway, 1882' (21 July 1882–1 September 1882), University of Aberdeen Museums and Special Collections (MS 3850/1/2)

— 'Micrococcus Poisoning' (Part One), *Journal of Anatomy and Physiology*, 16(Pt 4) (1882), 526–67

— 'Micrococcus Poisoning' (Part Two), *Journal of Anatomy and Physiology*, 17(Pt 1) (1882), 24–58

— 'On Flat-Foot and Its Cure by Operation', *Bristol Medico-Chirurgical Journal* 2(3) (1884), 1–20

— *Supplement to the Genealogical History of the Families of Ogston* (privately printed, 1897)

— 'Sketches of visit to South Africa, 1894–1895/Journal of visit to Russia, 1898–1899', University of Aberdeen Museums and Special Collections (MS 3850/1/4)

— 'The Effects of the Dum-dum Bullet from a Surgical Point of View', *The British Medical Journal*, 1(1952) (May 1898) 1425

— 'The Wounds Produced by Modern Small-Bore Bullets: The Dum-dum Bullet and the Soft-Nosed Mauser', *The British Medical Journal*, 2(1968) (September 1898)

— 'Continental Criticism of English Rifle Bullets', *The British Medical Journal*, 1(1995) (1899), 752–7

— 'The Peace Conference and The Dum-dum Bullet', *The British Medical Journal*, 2(2013) (1899), 278–81

— 'Address in Surgery: The Medical Services of The Army and Navy', *The British Medical Journal*, 2(2014) (1899), 337–45

— 'Boer War Journal (volume 1), 1899–1900', University of Aberdeen Museums and Special Collections (MS 3850/1/5)

— 'Boer War Journal (volume 2), 1900', University of Aberdeen Museums and Special Collections (MS 3850/1/6)

— 'A New Principle of Curing Club-Foot in Severe Cases in Children a Few Years Old', *The British Medical Journal*, 1(2164) (1902)

— 'First World War Journal (volume 1), 1915–1916', University of Aberdeen Museums and Special Collections (MS 3850/1/7)

— 'First World War Journal (volume 2), 1916–1917', University of Aberdeen Museums and Special Collections (MS 3850/1/8)
— 'Our Wounded – Sphagnum Moss as a Dressing', *The National Review*, 67(402) (August 1916), 870–5
— *Reminiscences of Three Campaigns* (London: Hodder and Stoughton, 1919)
— *The Prehistoric Antiquities of the Howe of Cromar* (Aberdeen: Aberdeen University Press, 1931)
Ogston, Walter, ed., *Sir Alexander Ogston K.C.V.O.* (Aberdeen: Aberdeen University Press, 1943)
— 'My Memoirs, Volume 1: 1873–1891', University of Aberdeen Museums and Special Collections (MS 3850/4/6/7/3/1)
Owen, Wilfred, 'Dulce Et Decorum Est', in *The Poems of Wilfred Owen*, ed. by Edmund Blunden (1946; rep. London: Chatto & Windus)
Pankhurst, E. Sylvia, *The Suffragette: The History of the Women's Militant Suffrage Movement*, 1905–1910 (New York: Sturgis & Walton, 1911)
Pennington, Carolyn, *The Modernisation of Medical Teaching at Aberdeen in the Nineteenth Century* (Aberdeen: Aberdeen University Press, 1994)
Raugh, Harold E., 'Bullets' in *The Victorians at War, 1815–1914: An Encyclopaedia of British Military History* (Santa Barbara, CA: ABC-CLIO, 2004)
Rennie, David A., 'The Real British Red Cross and Hemingway's A Farewell to Arms', *Hemingway Review* 37(2) (Spring 2018), 25–41
Reynolds, Michael, *Hemingway's First War: The Making of* A Farewell to Arms (Princeton: Princeton University Press, 1975)
Scotland, Tom, and Steven Heys, *Wars, Pestilence, and the Surgeon's Blade* (Solihull: Helion, 2013)
Scotland, Tom, and Ann Boyer, *Henry Gray: Surgeon in the Great War* (Edinburgh: Capercaillie Books, 2015)
Stark, Freya, *Traveller's Prelude* (London: John Murray, 1950)
Trevelyan, George Macaulay, *Scenes from Italy's War* (Boston: Houghton Mifflin Company, 1919)
Trevelyan, Laura, *A Very British Family: The Trevelyans and Their World* (London: I. B. Tauris & Co. Ltd, 2006)
Young, Geoffrey, *The Grace of Forgetting* (London: Country Life Ltd, 1953)
Zuberi, Dan, *Cleaning Up: How Hospital Outsourcing Is Hurting Workers and Endangering Patients* (Ithaca: ILR Press, 2013)

Index

EU Authorised Representative:

Easy Access System Europe Mustamäe tee 50, 10621 Tallinn, Estonia

gpsr.requests@easproject.com

Printed and bound by CPI Group (UK) Ltd, Croydon, CR0 4YY

25/07/2025

01924422-0018